M336
Mathematics and Computing: a third-level course

GROUPS & GEOMETRY

UNIT GE6
THREE-DIMENSIONAL LATTICES AND POLYHEDRA

Prepared for the course team by
Joe Rooney & Fred Holroyd

The Open University

This text forms part of an Open University third-level course.
The main printed materials for this course are as follows.

Block 1
Unit IB1 Tilings
Unit IB2 Groups: properties and examples
Unit IB3 Frieze patterns
Unit IB4 Groups: axioms and their consequences

Block 2
Unit GR1 Properties of the integers
Unit GR2 Abelian and cyclic groups
Unit GE1 Counting with groups
Unit GE2 Periodic and transitive tilings

Block 3
Unit GR3 Decomposition of Abelian groups
Unit GR4 Finite groups 1
Unit GE3 Two-dimensional lattices
Unit GE4 Wallpaper patterns

Block 4
Unit GR5 Sylow's theorems
Unit GR6 Finite groups 2
Unit GE5 Groups and solids in three dimensions
Unit GE6 Three-dimensional lattices and polyhedra

The course was produced by the following team:

Andrew Adamyk (BBC Producer)
David Asche (Author, Software and Video)
Jenny Chalmers (Publishing Editor)
Bob Coates (Author)
Sarah Crompton (Graphic Designer)
David Crowe (Author and Video)
Margaret Crowe (Course Manager)
Alison George (Graphic Artist)
Derek Goldrei (Groups Exercises and Assessment)
Fred Holroyd (Chair, Author, Video and Academic Editor)
Jack Koumi (BBC Producer)
Tim Lister (Geometry Exercises and Assessment)
Roger Lowry (Publishing Editor)
Bob Margolis (Author)
Roy Nelson (Author and Video)
Joe Rooney (Author and Video)
Peter Strain-Clark (Author and Video)
Pip Surgey (BBC Producer)

With valuable assistance from:

Maths Faculty Course Materials Production Unit
Christine Bestavachvili (Video Presenter)
Ian Brodie (Reader)
Andrew Brown (Reader)
Judith Daniels (Video Presenter)
Kathleen Gilmartin (Video Presenter)
Liz Scott (Reader)
Heidi Wilson (Reader)
Robin Wilson (Reader)

The external assessor was:
Norman Biggs (Professor of Mathematics, LSE)

The Open University, Walton Hall, Milton Keynes, MK7 6AA.

First published 1994. Reprinted 1997, 2002, 2007.

Copyright © 1994 The Open University

All rights reserved. No part of this publication may be reproduced, stored in a retrieval system or transmitted in any form or by any means, without written permission from the publisher or a licence from the Copyright Licensing Agency Limited. Details of such licences (for reprographic reproduction) may be obtained from the Copyright Licensing Agency Ltd of 90 Tottenham Court Road, London, W1P 9HE.

Edited, designed and typeset by the Open University using the Open University TEX System.

Printed in Malta by Gutenberg Press Limited.

ISBN 0 7492 2174 7

This text forms part of an Open University Third Level Course. If you would like a copy of *Studying with the Open University*, please write to the Central Enquiry Service, PO Box 200, The Open University, Walton Hall, Milton Keynes, MK7 6YZ. If you have not already enrolled on the Course and would like to buy this or other Open University material, please write to Open University Educational Enterprises Ltd, 12 Cofferidge Close, Stony Stratford, Milton Keynes, MK11 1BY, United Kingdom.

1.3

The paper used for this book is FSC-certified and totally chlorine-free. FSC (the Forest Stewardship Council) is an international network to promote responsible management of the world's forests.

CONTENTS

Study guide		4
Introduction		5
1	**Space lattices**	**6**
	1.1 Lattices and crystals	6
	1.2 The basic parallelepiped	9
	1.3 Lattices in layers	11
	1.4 The Lattice Card and its overlays	13
2	**The seven crystal systems**	**15**
	2.1 Introduction	15
	2.2 Space lattices with zero offset	15
	2.3 Space lattices with non-zero offset	20
3	**The Bravais lattices**	**23**
	3.1 The base-centred monoclinic lattice	23
	3.2 Body-centred space lattices	24
	3.3 Face-centred space lattices	26
4	**Polyhedra**	**28**
	4.1 The semi-regular polyhedra	28
	4.2 Space-filling polyhedra (video subsection)	33
	4.3 Counting with groups revisited	33
5	**Conclusion**	**35**
	5.1 Looking back	35
	5.2 Looking forward	36
	5.3 Revision exercises	37
Solutions to the exercises		39
Objectives		45
Index		46

STUDY GUIDE

As the final unit of the Geometry stream, this unit is intended as a fairly gentle introduction to three-dimensional versions of concepts that you have met at some length in the plane.

You should find Section 1 quite straightforward. The next three sections are somewhat more demanding, not so much in terms of mathematical concepts or techniques, but more in terms of the demands made on your ability to visualize in three dimensions.

In order to help you, we have provided a Lattice Card with overlays, and have suggested several experiments in which you stack these overlays in order to simulate the third dimension. You will find the card and these overlays in your *Geometry Envelope*, and you will need them in your study of Sections 2 and 3.

In the final section, there is a brief review of the Geometry stream of the course and then a look forward to ideas which you may encounter if you read further in this area. There is also a set of exercises to help you to revise the earlier Geometry units in preparation for the examination.

The video programme associated with this unit is VC4B. It is best viewed after you have studied the unit up to and including Subsection 4.1.

There is no audio programme associated with this unit.

INTRODUCTION

In *Unit GE3* you were introduced to the idea of a lattice in two-dimensional space. This is defined as a set of points in the plane which is generated by integer combinations of a pair of linearly independent vectors. The pair of vectors is referred to as a *basis* of the lattice.

In Section 1, we extend the concept of a lattice of points into three-dimensional space. Naturally this three-dimensional array of points will be generated by integer combinations of three linearly independent vectors.

In Sections 2 and 3, we show that in three dimensions (as in two) there is only a limited number of possible types of lattice — actually fourteen possibilities. These have become known as the *Bravais lattices*, named after the crystallographer Auguste Bravais who, in 1850, classified the possible types of three-dimensional lattice. He determined that a three-dimensional lattice can have one of just seven point groups, giving rise to the seven so-called *crystal systems*. (Some of these are associated with more than one geometric type of lattice, bringing the total number of lattice types to fourteen.)

In Section 4, we investigate the three-dimensional analogues of polygons. These are *polyhedra*, which you are already familiar with from several examples such as the five regular solids. Various other types of polyhedra with interesting symmetry properties can be obtained from the regular solids by chopping away the vertices, as you will see. The video programme considers the three-dimensional analogue of the problem of constructing tilings of the plane using polygons. This corresponds in three dimensions to filling space with polyhedra — a technique used by architects, both human and insect!

No new concepts are introduced in Section 5. It briefly reviews the results you have seen in the Geometry stream of the course, and tells you a little about what is known of their higher-dimensional analogues, before presenting a set of exercises specially designed to help you to prepare for the examination.

1 SPACE LATTICES

1.1 Lattices and crystals

In this course so far, we have seen geometric objects with one-dimensional and two-dimensional periodicity. *Friezes* have one-dimensional periodicity, while *plane lattices* and *wallpaper patterns* have two-dimensional periodicity.

One way to describe the 'dimensional periodicity' of a geometric object is in terms of translational symmetries of an *associated lattice*; thus, at the beginning of *Unit GE4*, we defined a (plane) lattice L to be *associated* with a wallpaper pattern W if it possesses the same translational symmetries. We did not introduce frieze patterns in *Unit IB3* in quite this way, but it should be clear that a frieze pattern F will have an associated lattice $L(\mathbf{a})$ which is one-dimensional.

There are few familiar examples of objects with three-dimensional periodicity. But the situation is entirely different when we go beyond everyday experience and consider the scientific world. In particular, the fields of materials science and geology are rich in examples of *crystal structures* — the archetypal three-dimensional periodic structures. Some examples are shown in Figure 1.1.

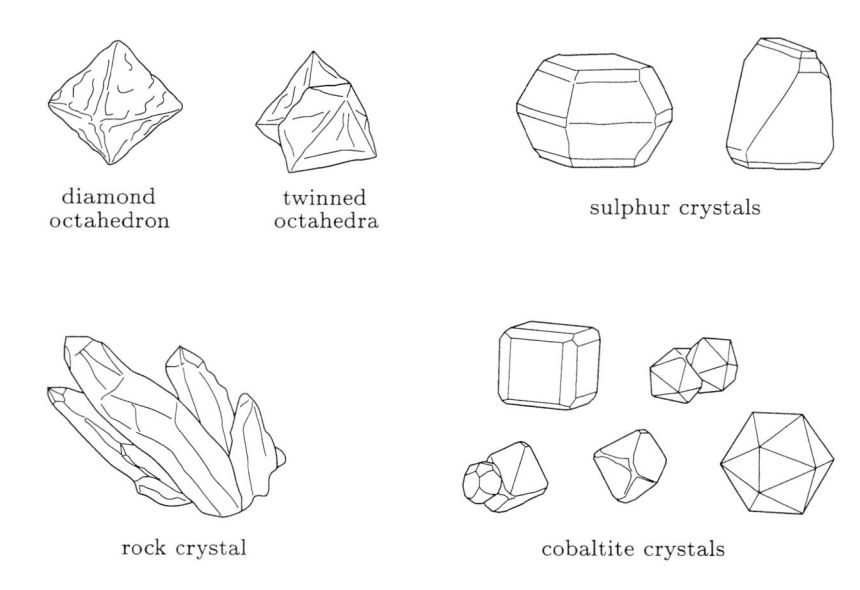

diamond octahedron

twinned octahedra

sulphur crystals

rock crystal

cobaltite crystals

Figure 1.1

First, we define a three-dimensional lattice as follows.

Definition 1.1 Three-dimensional (or space) lattice

A **three-dimensional** (or **space**) **lattice** is a set consisting of all those points in space whose position vectors constitute the set

$$L(\mathbf{a}, \mathbf{b}, \mathbf{c}) = \{n\mathbf{a} + m\mathbf{b} + r\mathbf{c} : n, m, r \in \mathbb{Z}\}$$

where $\mathbf{a}$, $\mathbf{b}$ and $\mathbf{c}$ are linearly independent vectors of $\mathbb{R}^3$.

Throughout this unit, space means three-dimensional space, modelled by $\mathbb{R}^3$.

Then we define a space lattice L to be *associated* with a pattern C in space if it possesses the same translational symmetries.

> **Definition 1.2 Crystal pattern**
>
> A **crystal pattern** C is a subset of space whose translational symmetries are those of some space lattice. Any such space lattice L is said to be **associated** with C.

Example 1.1

Consider a set of building blocks in the shape of cubes (all of the same size), some dark and some light. It would be possible to build a crystal pattern in layers, each layer consisting of blocks arranged in chessboard fashion, with successive layers arranged so that each block is directly over a block of the other colour, as shown in Figure 1.2.

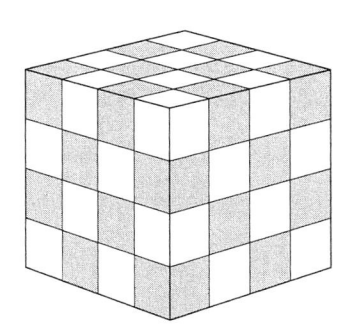

Figure 1.2

Regarding the set of points of space belonging to the dark blocks as defining the pattern, there is clearly three-dimensional periodicity, so we have a crystal pattern. ♦

The dark blocks correspond in $\mathbb{R}^3$ to the 'inked subset' of $\mathbb{R}^2$, with which we formalized the concept of a wallpaper pattern in *Unit GE4*.

Exercise 1.1

If each side of a block has length 1 unit, and the sides are parallel to the x-, y- and z-axes, describe a space lattice associated with the above crystal structure. That is, give vectors $\mathbf{a}, \mathbf{b}, \mathbf{c}$ such that $L(\mathbf{a}, \mathbf{b}, \mathbf{c})$ is the associated lattice based at the origin.

In the nineteenth century, considerable progress was made in understanding the structure and properties of matter, especially in determining the relationship between microscopic structure and macroscopic properties. Many substances were known to be crystalline, but the underlying nature of crystals was not well understood. One of the many researchers studying this subject was the crystallographer Auguste Bravais; and, as a result of his work, a classification system for crystal structures was eventually derived.

The number of different possibilities is much greater than in the plane — there are 230 different *crystallographic groups* (compared with just seventeen wallpaper groups), and fourteen different types of space lattice, the **Bravais lattices** (compared with the five plane lattice types).

The shapes of crystals arise from the symmetry of the spatial arrangement of the constituent atoms and molecules forming the substance of the crystal. This spatial arrangement depends partly on the size and shape of the molecules and on how closely they can be 'packed' together. So, what types of close-packed arrangements are possible?

Another number that you may sometimes hear quoted as the number of 'different crystal types' is 32. This is the number of different *point groups* that a crystal pattern may have (compared with ten possibilities for wallpaper patterns).

Exercise 1.2

Consider a two-dimensional 'crystal' constructed from atoms of just one type. Model the atoms as equally sized circular discs and hence determine the most closely packed arrangement of these circular atoms.

In three-dimensional space we can perform the same exercise if we model individual atoms as spheres and consider how these spheres can be packed closely together. But, as you will discover when you view the video programme for this unit, there is now an infinite number of different possible spatial arrangements which are equally closely packed. Of these, there are two that give rise to crystal structures in a straightforward way. These are shown in Figure 1.3. The two associated lattices, the *face-centred cubic lattice* and the *hexagonal lattice*, are among the fourteen that we shall examine in the remaining sections.

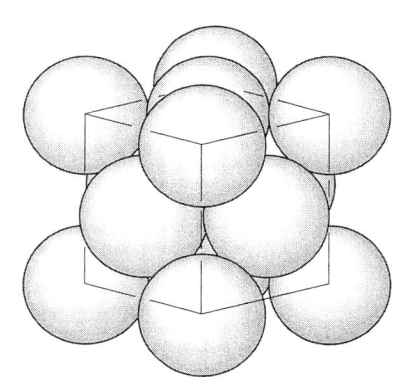

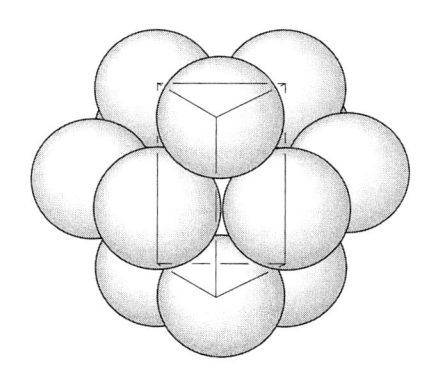

face-centred cubic   hexagonal

Figure 1.3

The fact that, of all space packings *whose centres lie on a space lattice*, these two give the closest packing, was proved by Karl Friedrich Gauss in 1831. It is surprisingly difficult to show that there is no *more general* sphere packing that gives closer packing. Wu-Yi Hsiang presented a paper of nearly one hundred pages in 1993, claiming to prove this; at the time of going to press the jury is still out!

They differ in their symmetry. Both types occur naturally in crystal structures, and which particular one arises for a given substance depends on extra physical factors in addition to the geometric constraints. Thus, the atoms of metals such as copper and silver form face-centred cubic crystals, whereas those of metals such as magnesium and beryllium form hexagonal crystals. In general, for metals, the physical factor determining which structure occurs is the energy content of the pool of 'free' electrons which provide their relatively high electrical conductivity. Similarly, for many other crystalline substances a major physical factor determining the structure is the slightly non-isotropic nature of interatomic forces (in other words, the force between adjacent atoms depends to some extent on the direction in space of the line joining their centres).

Thus the geometry of close-packing is not the only factor in determining the spatial arrangement of atoms or molecules in a crystal structure. Indeed, many substances exhibit more than one such arrangement depending on the prevailing physical conditions. Thus, for example, iron has a face-centred cubic lattice at relatively high temperatures but a *body-centred cubic lattice* at low temperatures (see Figure 1.4).

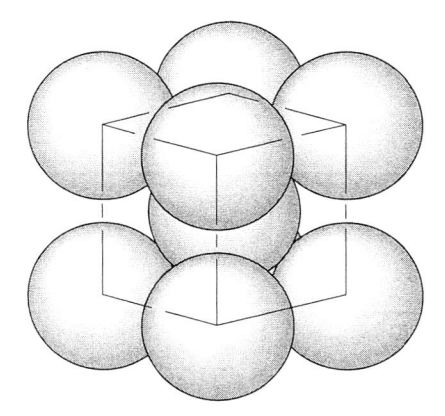

body-centred cubic

Figure 1.4

1.2 The basic parallelepiped

Much of the mathematics of point lattices is the same regardless of the dimensionality of the space. In particular, the concept of a basis of a lattice is essentially the same whatever the dimension — in space, a set of three vectors $\{\mathbf{a}, \mathbf{b}, \mathbf{c}\}$ is a **basis** for a lattice L if the three are linearly independent and

$$L = L(\mathbf{a}, \mathbf{b}, \mathbf{c}) = \{n\mathbf{a} + m\mathbf{b} + r\mathbf{c} : n, m, r \in \mathbb{Z}\}.$$

Theorems 1.1 and 1.2 of *Unit GE3* carry over except that, in three dimensions, 'triangle' must be replaced by 'tetrahedron'. We shall not prove these results, but merely state them.

Theorem 1.1

A linearly independent set $\{\mathbf{a}, \mathbf{b}, \mathbf{c}\}$ of three vectors in a space lattice L is a basis for L if and only if the corresponding tetrahedron $OABC$ contains no lattice points other than O, A, B and C.

Theorem 1.2 Identity of lattices

Two space lattices $L(\mathbf{a}, \mathbf{b}, \mathbf{c})$ and $L(\mathbf{a}', \mathbf{b}', \mathbf{c}')$ are identical if and only if the transition matrix from one basis to the other has integer entries and its determinant is equal to 1 or -1.

The concept of *basic parallelogram* also has a spatial analogue, of course. The solid figure corresponding to a parallelogram has six faces, all parallelograms, opposite pairs of faces being parallel. It is called a *parallelepiped* (see Figure 1.5).

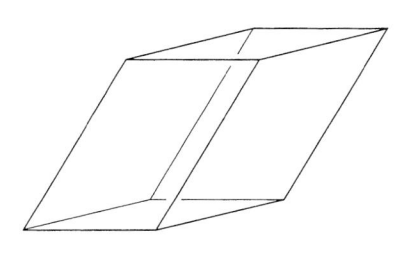

Figure 1.5

In just the same way that any parallelogram can be obtained from the unit square by a suitable affine transformation, so any parallelepiped can be obtained from the unit cube by an affine transformation.

> **Definition 1.3 Basic parallelepiped**
>
> If $\{\mathbf{a}, \mathbf{b}, \mathbf{c}\}$ is a basis for a space lattice L, then the vectors $\mathbf{0}, \mathbf{a}, \mathbf{b}, \mathbf{c}, \mathbf{a}+\mathbf{b}, \mathbf{a}+\mathbf{c}, \mathbf{b}+\mathbf{c}$ and $\mathbf{a}+\mathbf{b}+\mathbf{c}$ form the corners of a parallelepiped known as a **basic parallelepiped** for L.

Just as in the plane, there are many different possible choices (infinitely many, in fact) of basis for a space lattice, giving rise to an infinity of different basic parallelepipeds. However, they all have the same volume.

How is the volume of a parallelepiped calculated? Again, the answer is a direct analogue of the case for the plane. Recall that, in the plane, if $\mathbf{a} = (a_1, a_2)$ and $\mathbf{b} = (b_1, b_2)$, then the area of the corresponding parallelogram is $|a_1 b_2 - a_2 b_1|$. This is, of course, the modulus of the determinant

$$\begin{vmatrix} a_1 & a_2 \\ b_1 & b_2 \end{vmatrix}.$$

In the three-dimensional case, if $\mathbf{a} = (a_1, a_2, a_3), \mathbf{b} = (b_1, b_2, b_3)$ and $\mathbf{c} = (c_1, c_2, c_3)$, then the volume of the corresponding parallelepiped is the modulus of the 3×3 determinant

$$\begin{vmatrix} a_1 & a_2 & a_3 \\ b_1 & b_2 & b_3 \\ c_1 & c_2 & c_3 \end{vmatrix}.$$

Exercise 1.3

Find the volumes of the parallelepipeds defined by the following triples of vectors.

(a) $\mathbf{a} = (1, 0, 0), \mathbf{b} = (1, 1, 0), \mathbf{c} = (1, 2, 3)$
(b) $\mathbf{a} = (1, 0, 0), \mathbf{b} = (1, 1, 1), \mathbf{c} = (1, 2, 3)$
(c) $\mathbf{a} = (1, 1, 1), \mathbf{b} = \left(\frac{1}{2}, 1, 2\right), \mathbf{c} = (1, 3, 9)$
(d) $\mathbf{a} = (1, 1, 0), \mathbf{b} = (1, -1, 0), \mathbf{c} = (1, 2, 3)$

Continuing the analogy with plane lattices, we have the following result.

> **Theorem 1.3**
>
> Let $L = L(\mathbf{a}, \mathbf{b}, \mathbf{c})$ be a space lattice, and let $\mathbf{a}', \mathbf{b}', \mathbf{c}'$ be the position vectors of three points of L. Then $\{\mathbf{a}', \mathbf{b}', \mathbf{c}'\}$ is a basis of L if and only if the volume of the corresponding parallelepiped is the same as that of the basic parallelepiped defined by $\{\mathbf{a}, \mathbf{b}, \mathbf{c}\}$.

The proof is essentially the same as that for Theorem 1.4 of *Unit GE3*.

Exercise 1.4

Let $L = L((1, 0, 0), (0, 1, 0), (0, 0, 1))$. Which of the sets of vectors in Exercise 1.3 are also a basis for L?

1.3 Lattices in layers

It is much more difficult to visualize accurately in three dimensions than in two, and consequently a good way to deal with space lattices in practice is to consider them to be built up, layer by layer, out of plane lattices.

By way of introducing this technique, it is worth recalling that we did something analogous in Subsection 1.2 of *Unit GE3*. We considered a plane lattice, with a basis $\{\mathbf{a}, \mathbf{b}\}$, and noted that every lattice point is therefore represented by a vector $n\mathbf{a} + m\mathbf{b}$, where n and m are integers. We then considered the lines of points obtained by letting n vary while keeping m fixed (see Figure 1.6). That is to say, for each fixed m in $\mathbb{Z}$, we considered the line

$$\mathcal{L}_m = \{n\mathbf{a} + m\mathbf{b} : n \in \mathbb{Z}\}.$$

The plane lattice consists of a discrete set of points, but we draw a continuous straight line through each subset of points $\mathcal{L}_m$ to show how these points are aligned in a row.

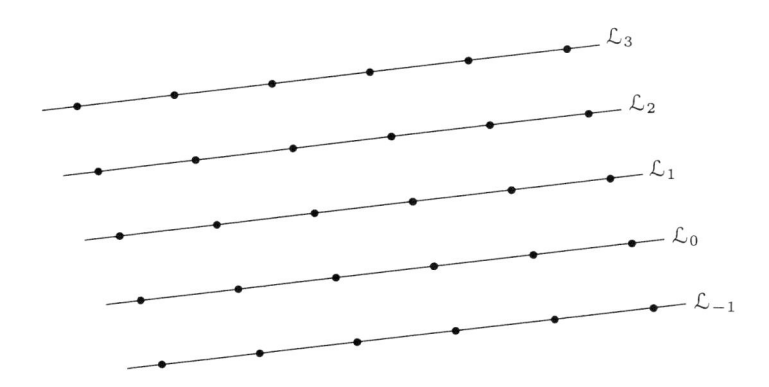

Figure 1.6

For the moment, let us simplify matters by choosing a coordinate system in which $\mathbf{a}$ is the vector $(1,0)$, so that $\mathcal{L}_0$ is the 'one-dimensional lattice' consisting of the integer points of $\mathbb{R}^1$, i.e. of $\mathbb{R}$ (see Figure 1.7).

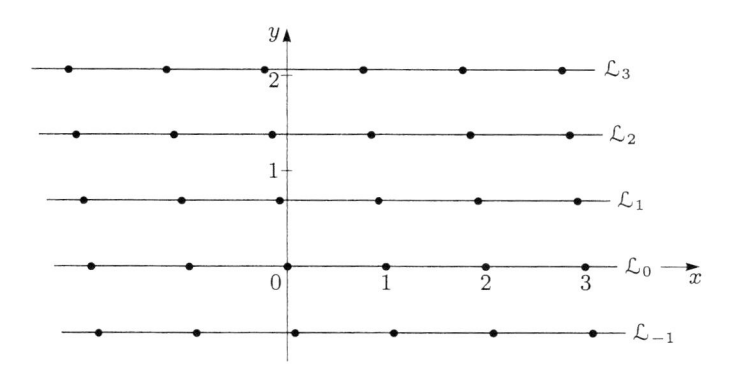

Figure 1.7

Now, given that L has been split up into lines as in Figure 1.7, indicating a choice of $\mathbf{a}$ as the first basis vector, what possible choices are there for the second basis vector, $\mathbf{b} = (b_1, b_2)$?

Clearly, we must choose $\mathbf{b}$ to be a lattice point on one of the lines $\mathcal{L}_1, \mathcal{L}_{-1}$. If we were to choose $\mathbf{b}$ to lie on $\mathcal{L}_2$, for example, then the integer combinations of $\mathbf{a}$ and $\mathbf{b}$ would consist only of lattice points on the even numbered lines, and not of the whole lattice. On the other hand, *any* of the lattice points on $\mathcal{L}_1$ or $\mathcal{L}_{-1}$ will do, since adding multiples of $\mathbf{a}$ will then generate the whole of that line, while choosing different multiples of $\mathbf{b}$ will give every line.

Notice that we are assuming that L has already been 'given' to us; we are not constructing it from scratch. (If we were, then of course $\mathbf{b}$ could be any vector whatever as long as it was linearly independent of $\mathbf{a}$.)

Thus, the y-coordinate b_2 of **b** may be chosen to be positive and equal to the vertical separation of the lines, while the x-coordinate b_1 may be chosen modulo 1, and may therefore be chosen to be zero if the points of $\mathcal{L}_1$ lie vertically above those of $\mathcal{L}_0$, and between 0 and 1 otherwise. This value of b_1 modulo 1 is a quantity which we shall call the *offset* of **b** relative to **a**.

Exercise 1.5

Let $\mathbf{a} = (1, 0)$. For each of the following choices of **b**, find the offset of **b** relative to **a**.

(a) $\mathbf{b} = (0, 2)$
(b) $\mathbf{b} = (1, 1)$
(c) $\mathbf{b} = \left(\frac{1}{2}, \frac{3}{2}\right)$
(d) $\mathbf{b} = \left(\frac{9}{10}, 2\right)$
(e) $\mathbf{b} = \left(\frac{14}{3}, 3\right)$

More generally, given any two linearly independent vectors **a** and **b**, we may express **b** as

$$\mathbf{b} = \mathbf{p} + (n + \omega)\mathbf{a}$$

where **p** is orthogonal to **a**, n is an integer, and $0 \leq \omega < 1$.

Exercise 1.6

Find the value of ω when:

(a) $\mathbf{a} = (2, 0), \mathbf{b} = (1, 1)$;
(b) $\mathbf{a} = (0, 1), \mathbf{b} = \left(\frac{1}{2}, \frac{1}{3}\right)$;
(c) $\mathbf{a} = (1, 1), \mathbf{b} = (0, 2)$;
(d) $\mathbf{a} = (1, 1), \mathbf{b} = (0, 3)$.

> **Definition 1.4 Offset (two vectors)**
>
> Let **a** and **b** be any two linearly independent vectors. Then, expressing **b** as
>
> $$\mathbf{b} = \mathbf{p} + (n + \omega)\mathbf{a}$$
>
> where **p** is orthogonal to **a**, $n \in \mathbb{Z}$ and $0 \leq \omega < 1$, the value ω is the **offset** of **b** relative to **a**.

When there is no risk of confusion, we shall omit the phrase 'relative to **a**'.

The importance of offset is that it gives a means of identifying the type of a plane lattice, in a way that extends to three dimensions.

Exercise 1.7

(a) Show that a plane lattice L is square or rectangular if and only if it has a basis $\{\mathbf{a}, \mathbf{b}\}$ for which **b** has zero offset.
(b) Let **b** have offset $\frac{1}{2}$; more particularly, assume that $\mathbf{b} = \mathbf{p} + \frac{1}{2}\mathbf{a}$ where **p** is orthogonal to **a** (i.e. assume that $n = 0$ in the definition of offset). Show that $L(\mathbf{a}, \mathbf{b})$ has all the symmetries of a rhombic lattice — that is, it is rhombic, square or hexagonal.

Hint Consider the basis $\{\mathbf{b}, \mathbf{b} - \mathbf{a}\}$, and calculate $\mathbf{b} \cdot \mathbf{b}$ and $(\mathbf{b} - \mathbf{a}) \cdot (\mathbf{b} - \mathbf{a})$.

The fact that a rhombic lattice has a basis $\{\mathbf{a}, \mathbf{b}\}$ in which $\mathbf{b}$ has offset $\frac{1}{2}$ is connected with the name 'centred rectangular' which is sometimes given to a rhombic lattice. If $\{\mathbf{a}, \mathbf{b}\}$ is such a basis for L, then the vector $2\mathbf{b}$ has zero offset, and so the lattice L' formed from alternate lines $\mathcal{L}_0, \mathcal{L}_2, \mathcal{L}_4, \ldots$ is a rectangular lattice (see Figure 1.8).

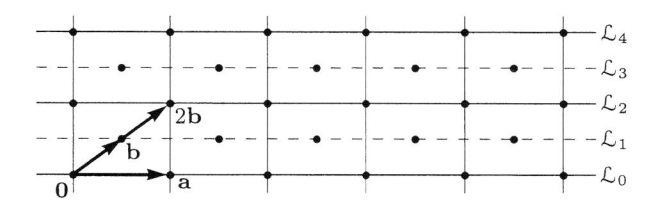

Figure 1.8

The odd-numbered lines of L contribute lattice points at the centres of the rectangles of L'.

It is now time to return to the subject of space lattices.

In just the same way that we found it convenient to start thinking about plane lattices by taking the basis vector $\mathbf{a}$ to lie along the x-axis, so, in thinking about space lattices, we shall find it useful to choose our coordinate system for a space lattice $L(\mathbf{a}, \mathbf{b}, \mathbf{c})$ so that $\mathbf{a}$ and $\mathbf{b}$ lie in the xy-plane. This means that the points of $L(\mathbf{a}, \mathbf{b}, \mathbf{c})$ that lie in the xy-plane are exactly the points of the plane lattice $L(\mathbf{a}, \mathbf{b})$. We may then think of $L(\mathbf{a}, \mathbf{b}, \mathbf{c})$ as an infinite stack of copies of $L(\mathbf{a}, \mathbf{b})$. Thus, for each fixed integer r, the set of lattice points of the form $n\mathbf{a} + m\mathbf{b} + r\mathbf{c}$ $(n, m \in \mathbb{Z})$ can be obtained by translating the whole of $L(\mathbf{a}, \mathbf{b})$ by the vector $r\mathbf{c}$. For each r, let us denote this translated copy by L_r (see Figure 1.9, where L_0 and L_3 are shaded).

The space lattice consists of a discrete set of points, but we draw a two-dimensional grid of continuous straight lines through each subset of points L_r to show how these points are aligned in rows and columns.

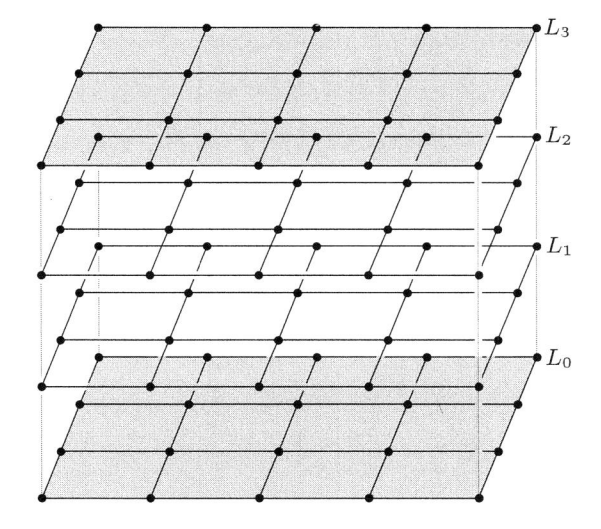

Figure 1.9

1.4 The Lattice Card and its overlays

The Lattice Card and its overlays in the *Geometry Envelope* are designed to help you to visualize the stacking of layers of plane lattices to form a space lattice, and to experiment with the rotational symmetries that occur with different offsets.

The card has plane lattices of the five types printed on it (three on Side 1 and two on Side 2), together with certain rotation centres. There are also three overlays for each side: Overlay 1 printed in red, Overlay 2 in blue and Overlay 3 in green. The idea is to stack them, then insert a pin through the overlays and rotate them bodily, keeping the card fixed so that you can check which rotations bring the lattice points on the overlays back to their original positions. There is no need to use them if you are happy without them, but we think you will find them helpful.

Lay the Lattice Card down in front of you, with Side 1 uppermost. The first lattice on the card is a parallelogram lattice, with the parallelograms corresponding to a particular basis $\{\mathbf{a}, \mathbf{b}\}$ drawn in. The large dots at the corners of the basic parallelograms are the lattice points; the other markings correspond to the orbits of 2-centres, and you should ignore these for the present.

In most of the experiments you will be asked to perform with the Lattice Card and its overlays, the card itself is not *meant to be part of the space lattice — it is just there to mark a starting position. The one exception is Experiment 2.7 on page 22.*

Now superimpose the red overlay for Side 1 precisely over the card. Regard the parallelogram lattice on this overlay as L_0. Next, hold the blue overlay for Side 1 over the card, to represent L_1. The basis vector $\mathbf{c} = (c_1, c_2, c_3)$ of the three-dimensional lattice you are constructing is determined by exactly where you hold this overlay; the value of c_3 can be varied by raising or lowering the overlay, while c_1 and c_2 can be varied by translating it horizontally (that is, in the x- and y-directions).

In practice, you will probably find it satisfactory to let the overlay lie on the card, shifted to represent (c_1, c_2), and just imagine c_3; but if you wish to, you may like to devise your own way of supporting the blue overlay above the plane of the card. (Blu-Tack or a similar product could be used — or perhaps you can be more ingenious.)

Which of the lattice points on the blue overlay represents the vector $\mathbf{c}$? Clearly it makes no difference — any one of them gives rise to the same space lattice $L(\mathbf{a}, \mathbf{b}, \mathbf{c})$. All that matters is the offset of $\mathbf{c}$ — but this time, the offset must be relative to *both* of the vectors $\mathbf{a}$ and $\mathbf{b}$. That is to say, we may express $\mathbf{c} = (c_1, c_2, c_3)$ as

$$\mathbf{c} = (0, 0, c_3) + (n + \lambda)\mathbf{a} + (m + \mu)\mathbf{b},$$

where $n, m \in \mathbb{Z}$, $0 \leq \lambda < 1$ and $0 \leq \mu < 1$. The pair (λ, μ) is the *offset* of $\mathbf{c}$ relative to $\{\mathbf{a}, \mathbf{b}\}$.

Exercise 1.8

Let $\mathbf{a} = (1, -1, 0), \mathbf{b} = (1, 2, 0), \mathbf{c} = (2, 3, 2)$. Find the offset of $\mathbf{c}$ relative to $\{\mathbf{a}, \mathbf{b}\}$.

Hint You will need to use simultaneous equations to express $(2, 3, 0)$ in terms of $\mathbf{a}$ and $\mathbf{b}$.

More generally, for any three linearly independent vectors $\mathbf{a}, \mathbf{b}, \mathbf{c}$ of $\mathbb{R}^3$, we may define the offset of $\mathbf{c}$ relative to $\{\mathbf{a}, \mathbf{b}\}$ much as we did in the plane case.

Definition 1.5 Offset (three vectors)

Let $\{\mathbf{a}, \mathbf{b}, \mathbf{c}\}$ be three linearly independent vectors. Then, expressing $\mathbf{c}$ as

$$\mathbf{c} = \mathbf{p} + (n + \lambda)\mathbf{a} + (m + \mu)\mathbf{b}$$

where $\mathbf{p}$ is orthogonal to $\mathbf{a}$ and $\mathbf{b}$, $n, m \in \mathbb{Z}$, $0 \leq \lambda < 1$ and $0 \leq \mu < 1$, the pair (λ, μ) is the **offset** of $\mathbf{c}$ relative to $\{\mathbf{a}, \mathbf{b}\}$.

Again, we shall normally omit the phrase 'relative to $\{\mathbf{a}, \mathbf{b}\}$'. Also, we shall nearly always choose $\mathbf{a}$ and $\mathbf{b}$ to be in the xy-plane — that is to say, we shall express $L(\mathbf{a}, \mathbf{b}, \mathbf{c})$ in *layer form*. This will make it easier to relate it to the Lattice Card and its overlays.

In the next two sections, we shall ask you to consider the effect of rotations. In preparation for this, we suggest that you cut up each overlay into individual plane lattices, so that you only rotate the lattice that you actually want in each case.

2 THE SEVEN CRYSTAL SYSTEMS

2.1 Introduction

In *Unit GE3*, you saw that plane lattices can be classified into five types in terms of their symmetry groups.

These five types are sometimes grouped into just four *plane crystal systems*, corresponding to the groups of symmetries that fix the origin — namely, C_2, D_2, D_4 and D_6.

Exercise 2.1

Which two classes of plane lattice belong to the same plane crystal system?

In three dimensions, the possible types of space lattice fall into fourteen classes in terms of their full symmetry groups and into seven *crystal systems* in terms of the groups of symmetries that fix the origin. Each one of these is, of course, one of the finite groups of three-dimensional isometries that you met in *Unit GE5*.

Now, just as in two dimensions, there is a non-trivial three-dimensional isometry which fixes the origin and is a symmetry of *every* space lattice.

In two dimensions, this isometry is rotation through π, which has the effect of mapping every vector $\mathbf{x}$ in $\mathbb{R}^2$ to $-\mathbf{x}$.

In three dimensions, the isometry in question is again the function

$$\mathbf{x} \to -\mathbf{x} \quad (\mathbf{x} \in \mathbb{R}^3),$$

which we denote by σ_O.

You saw this notation in *Unit GE5*, where this isometry was called *central inversion*.

The fact that a space lattice consists of *every* integer combination of a set of basis vectors shows that, just as in two dimensions, the isometry σ_O is a symmetry of *every* space lattice; the lattice point $n\mathbf{a} + m\mathbf{b} + r\mathbf{c}$ maps to $(-n)\mathbf{a} + (-m)\mathbf{b} + (-r)\mathbf{c}$, which is another lattice point.

As we saw in *Unit GE5*, σ_O is not a rotation in three dimensions; as its determinant is -1, it is an indirect isometry.

This means that, for any space lattice, the group of isometries fixing the origin contains σ_O.

From now on, we shall follow the convention of *Unit GE4* and use G to denote the full symmetry group of a space lattice and H to denote the subgroup of G that fixes O.

For lattices, this subgroup H is the same as the point group which you met in Section 2 of *Unit GE4*, so there is no harm in using the same notation. This fact follows from Theorem 2.2 of *Unit GE3*, whose proof is valid in any number of dimensions.

Theorem 2.1

Let L be a space lattice, with point group H. Then H is of the form

$$H = H^+ \cup \sigma_O H^+$$

where H^+ is the group of rotations of L that fix O.

Proof

This follows directly from Theorem 2.5 of *Unit GE5*, combined with the fact just observed: that H always contains the indirect isometry σ_O. ∎

2.2 Space lattices with zero offset

Suppose we form a space lattice by stacking plane lattices vertically above each other, with zero offset — that is, we construct a space lattice $L(\mathbf{a}, \mathbf{b}, \mathbf{c})$, in layer form, where

$$\mathbf{a} = (a_1, a_2, 0), \quad \mathbf{b} = (b_1, b_2, 0), \quad \mathbf{c} = (0, 0, c).$$

Depending on the type of the original plane lattice, we can obtain six (not five!) different space lattices, with five distinct point groups, as follows.

Case 1 $L(\mathbf{a}, \mathbf{b})$ is a parallelogram lattice

Let $L(\mathbf{a}, \mathbf{b})$ be a plane parallelogram lattice, so that the only non-trivial symmetry of $L(\mathbf{a}, \mathbf{b})$ that fixes the origin is the rotation $r[\pi]$ of $\mathbb{R}^2$. We then construct the corresponding space lattice $L(\mathbf{a}, \mathbf{b}, \mathbf{c})$, where $\mathbf{c} = (0, 0, c)$ (see Figure 2.1).

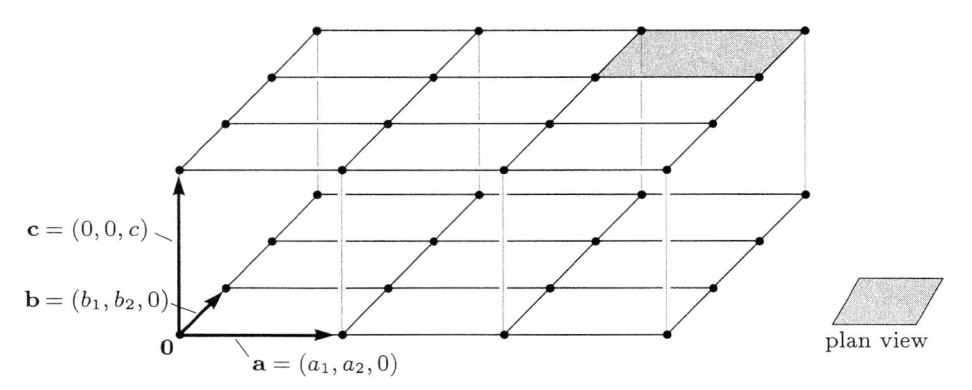

Figure 2.1

We have to be a bit careful with interpreting notation here. The $\mathbf{a}$ and $\mathbf{b}$ of $L(\mathbf{a}, \mathbf{b})$ are vectors in $\mathbb{R}^2$, whereas the $\mathbf{a}$, $\mathbf{b}$ and $\mathbf{c}$ of $L(\mathbf{a}, \mathbf{b}, \mathbf{c})$ are vectors in $\mathbb{R}^3$. Thus, there is an ambiguity between whether we think of $\mathbf{a}$ and $\mathbf{b}$ as (a_1, a_2) and (b_1, b_2) in $\mathbb{R}^2$, or as $(a_1, a_2, 0)$ and $(b_1, b_2, 0)$ in $\mathbb{R}^3$. Although it may perhaps be a little sloppy, we shall continue to use $\mathbf{a}$ and $\mathbf{b}$ in both cases, since the interpretation is always clear in practice.

Now, what isometry of $\mathbb{R}^3$ corresponds to the rotation $r[\pi]$ of $\mathbb{R}^2$? Clearly, the rotation through π about the z-axis in $\mathbb{R}^3$ has the desired effect, not only on the xy-plane itself, but also on all the planes parallel to the xy-plane. We shall denote this rotation by $R_z[\pi]$, where the capital letter is used to remind us that we are in three dimensions and the subscript z is used to describe the axis of rotation.

Since all the copies of $L(\mathbf{a}, \mathbf{b})$ are stacked vertically above one another in the space lattice $L(\mathbf{a}, \mathbf{b}, \mathbf{c})$, the rotation $R_z[\pi]$ must map the lattice points in each layer to lattice points in the same layer:

$$R_z[\pi](L_i) = L_i \qquad (i \in \mathbb{Z})$$

and so

$$R_z[\pi](L(\mathbf{a}, \mathbf{b}, \mathbf{c})) = L(\mathbf{a}, \mathbf{b}, \mathbf{c}).$$

This is shown in Figure 2.2.

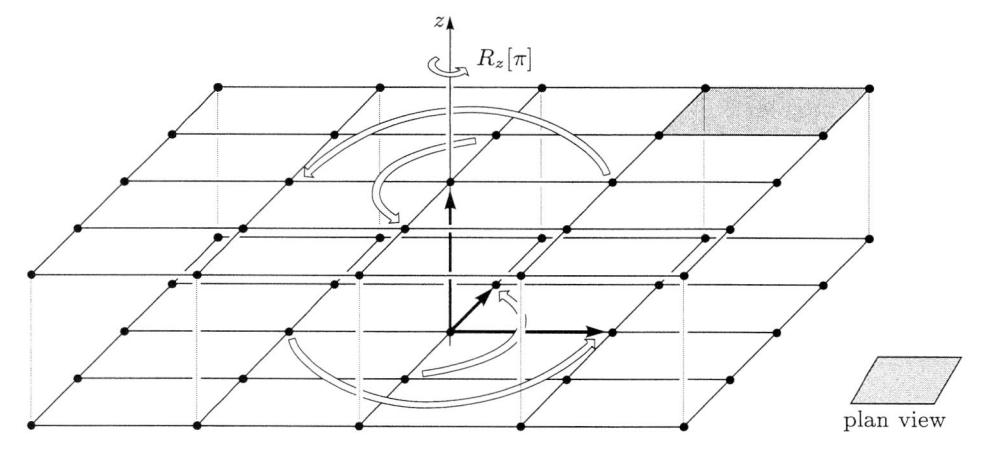

Figure 2.2

Experiment 2.1 Place both the red and the blue parallelogram lattice overlays directly over the parallelogram lattice on Side 1 of the Lattice Card, then place a pin through a lattice point of each overlay and a lattice point of the card. Regard the card as marking the original positions of the lattice points, and the two overlays as representing L_0 and L_1; then rotate both overlays through a half turn, using the pin as pivot. All the lattice points of both overlays end up over lattice points of the card, indicating that the rotation maps $L(\mathbf{a}, \mathbf{b}, \mathbf{c})$ to itself. ♦

Although we shall not prove it here, there is in fact no other non-trivial rotation about an axis through the origin that maps $L(\mathbf{a}, \mathbf{b}, \mathbf{c})$ to itself. In other words, in this case
$$H^+ = \{e, R_z[\pi]\},$$
so that
$$H = \{e, R_z[\pi], \sigma_O, \sigma_O R_z[\pi]\}.$$

In terms of Section 5 of *Unit GE5*, $H^+ = C_2$ and $H = C_2 \cup \sigma_O C_2$.

Exercise 2.2

Give a geometric description of the symmetry $\sigma_O R_z[\pi]$.

This lattice $L(\mathbf{a}, \mathbf{b}, \mathbf{c})$ is called a **primitive monoclinic lattice**, and any crystal pattern with an associated lattice such that (in a suitable coordinate system)
$$H = \{e, R_z[\pi], \sigma_O, \sigma_O R_z[\pi]\}$$
belongs to the **monoclinic crystal system**.

We shall see some non-primitive monoclinic lattices in Section 3.

The name *monoclinic* means 'one plane', and refers to the fact that the plane of $L(\mathbf{a}, \mathbf{b})$ has a special significance — it is the only plane through the origin having axes of rotational symmetry perpendicular to it.

Case 2 $L(\mathbf{a}, \mathbf{b})$ is a rectangular lattice

Let $L(\mathbf{a}, \mathbf{b})$ be a plane rectangular lattice. Let us choose the coordinate system so that $\mathbf{a}$ and $\mathbf{b}$ point in the x- and y-directions respectively.

Then, as well as e and $r[\pi]$, the group of symmetries of $L(\mathbf{a}, \mathbf{b})$ that fix the origin contains reflections in the x- and y-axes. These reflections of $\mathbb{R}^2$ can be obtained from isometries of $\mathbb{R}^3$ in two ways: either as rotations through π about these axes or as reflections in the vertical planes containing them. If $L(\mathbf{a}, \mathbf{b}, \mathbf{c})$ is then formed by stacking with zero offset, all these isometries of $\mathbb{R}^3$ will be symmetries of $L(\mathbf{a}, \mathbf{b}, \mathbf{c})$ (see Figure 2.3).

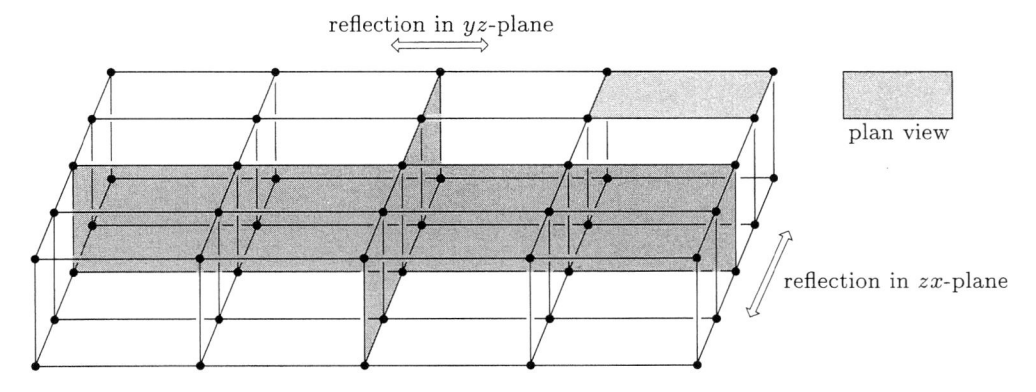

Figure 2.3

To be more specific, let us denote by $R_x[\pi]$ and $R_y[\pi]$ the three-dimensional rotations through π about the x- and y-axes, by Q_{zx} the reflection in the zx-plane and by Q_{yz} the reflection in the yz-plane. You saw in Exercise 2.2 that reflection in the xy-plane is also a symmetry of any lattice with zero offset; let us denote this reflection by Q_{xy}. In general, if $L(\mathbf{a}, \mathbf{b})$ is a rectangular lattice and $\mathbf{c}$ has zero offset, then the above symmetries constitute the whole of H:
$$H = \{e,\ R_x[\pi],\ R_y[\pi],\ R_z[\pi],\ \sigma_O,\ Q_{xy},\ Q_{yz},\ Q_{zx}\}.$$

Experiment 2.2 Place the red and blue rectangular lattice overlays precisely over the rectangular lattice on Side 1 of the Lattice Card. The rotation symmetry $R_z[\pi]$ is just as for the parallelogram case and, if you place your eye vertically over the card and overlays, you should be able to see that the vertical planes cutting along the x-axis and the y-axis are reflection planes.

If you now turn over the two overlays, you should be able to demonstrate the rotational symmetries $R_x[\pi]$ and $R_y[\pi]$. ♦

There will be special cases with more symmetry, for example if we choose **c** to be exactly the same length as **a** or **b**; but it would not be helpful to list these cases here.

Exercise 2.3

Write down the elements of H^+.

In the terminology of *Unit GE5*, H is the symmetry group of the digon DIH_2:

$$H = \Gamma(\text{DIH}_2).$$

The lattice $L(\mathbf{a}, \mathbf{b}, \mathbf{c})$ is in this case called a **primitive orthorhombic lattice**, and any crystal pattern with an associated lattice whose point group is $\Gamma(\text{DIH}_2)$ belongs to the **orthorhombic crystal system**.

We shall see a non-primitive orthorhombic lattice next.

Case 3 $L(\mathbf{a}, \mathbf{b})$ is a rhombic lattice

Let $L(\mathbf{a}, \mathbf{b})$ be a plane rhombic lattice. It is convenient in this case to choose **a** and **b**, and the coordinate system, so that the reflections in the x- and y-axes are again symmetries of $L(\mathbf{a}, \mathbf{b})$ — for example, $\mathbf{a} = (2, 1)$, $\mathbf{b} = (2, -1)$. Once again, choose **c** to have zero offset (and not to have length exactly equal to that of **a** or **b**) — so, for example, in three dimensions, we might have

$$\mathbf{a} = (2, 1, 0), \mathbf{b} = (2, -1, 0), \mathbf{c} = \left(0, 0, \tfrac{4}{3}\right).$$

In this case, the symmetries of $L(\mathbf{a}, \mathbf{b}, \mathbf{c})$ that fix the origin are exactly the same as for the stack of rectangular lattices:

$$H = \Gamma(\text{DIH}_2).$$

You may wish to check this by repeating Experiment 2.2, using the rhombic lattice on Side 1 of the Lattice Card and the corresponding red and blue overlays.

However, the whole symmetry group of L is not the same as in the case where we started with a rectangular lattice. The lattice is of a different type from that in which $L(\mathbf{a}, \mathbf{b})$ was rectangular; it is called a **base-centred orthorhombic lattice**.

The reason for this name is as follows. You have seen that a rhombic plane lattice can be obtained from a rectangular plane lattice by adding extra lattice points at the centre of each rectangle. Therefore, if we start with a primitive orthorhombic lattice, we may consider it as a vertical stack of rectangular lattices, and we may change it into a vertical stack of rhombic lattices by adding a lattice point to the centre of each rectangle lying in a horizontal plane (see Figure 2.4).

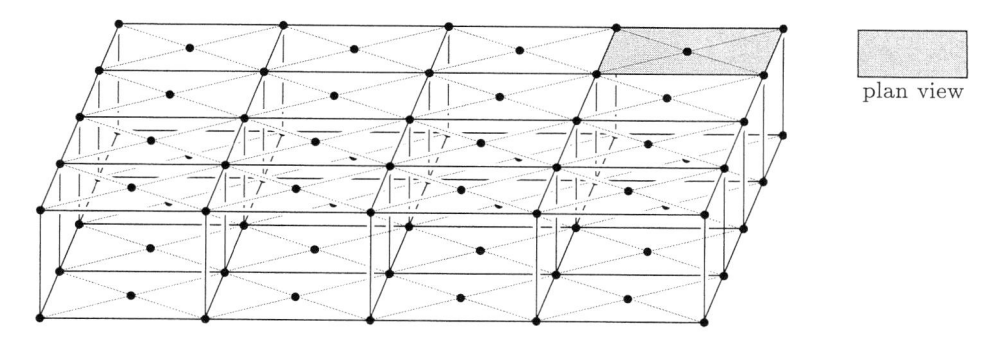

Figure 2.4

We can consider the vertices of the primitive orthorhombic lattice as the corners of a set of rectangular boxes, which are basic parallelepipeds that fill $\mathbb{R}^3$. Then we may regard the bottom face of each of these boxes as its 'base'. Hence, placing a new vertex at the centre of each 'base face' of each box converts the primitive orthorhombic lattice into a new lattice, an obvious name for which is 'base-centred orthorhombic', and which is in fact a vertical stack of rhombic lattices.

Admittedly there is an awkwardness of language here, in that the meaning of the word *base* has nothing to do with the concepts of *basic parallelogram*, *basic parallelepiped*, or *basis*; but it is the standard terminology.

Case 4 $L(\mathbf{a}, \mathbf{b})$ is a square lattice

Let $L(\mathbf{a}, \mathbf{b})$ be a plane square lattice, with $\mathbf{a}$ and $\mathbf{b}$ in the directions of the x- and y-axes respectively (see Figure 2.5).

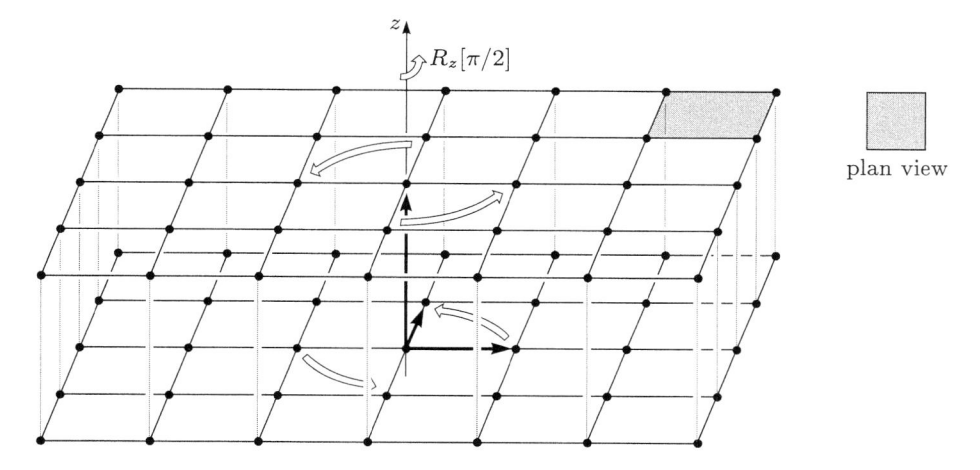

Figure 2.5

If we choose $\mathbf{c} = (0, 0, c)$ so that $||\mathbf{c}|| \neq ||\mathbf{a}||$, we obtain a lattice $L(\mathbf{a}, \mathbf{b}, \mathbf{c})$ whose group of symmetries fixing the origin includes rotation through $\pi/2$ about the z-axis. In fact, we have

$$H = \Gamma(\mathrm{DIH}_4).$$

Experiment 2.3 Repeat Experiment 2.2, this time with the square lattice on Side 2 of the Lattice Card and the corresponding red and blue overlays. You should, of course, see all the reflection and rotation symmetries that you saw before; but you should also find that rotations through $\pi/2$ are symmetries of the space lattice $L(\mathbf{a}, \mathbf{b}, \mathbf{c})$, as are reflections in vertical planes cutting the xy-plane diagonally. ♦

The lattice $L(\mathbf{a}, \mathbf{b}, \mathbf{c})$ is in this case called a **primitive tetragonal lattice**, and any crystal pattern with an associated lattice whose point group is $\Gamma(\mathrm{DIH}_4)$ belongs to the **tetragonal crystal system**.

We shall see some non-primitive tetragonal lattices in Section 3.

So far, the point groups of all our crystal patterns have had the useful property that there is a definite direction (which we took to be that of the z-axis) such that *every* axis of rotational symmetry is either in that direction or perpendicular to it — in other words, every point group so far has been the symmetry group of a dihedron or a subgroup thereof. But if we stick with the square lattice for $L(\mathbf{a}, \mathbf{b})$, and now let $\mathbf{c}$ be the same length as $\mathbf{a}$, for example, $\{\mathbf{a}, \mathbf{b}, \mathbf{c}\} = \{(1,0,0), (0,1,0), (0,0,1)\}$, then we get a point group with rotation axes in considerably more directions.

Exercise 2.4

What is the point group of the lattice $L((1,0,0), (0,1,0), (0,0,1))$? In how many directions are there axes of rotational symmetry?

This lattice is (not surprisingly) called a **primitive cubic lattice**, and any crystal pattern with an associated lattice whose point group is $\Gamma(\mathrm{CUBE})$ belongs to the **cubic crystal system**. Although this lattice is intuitively easy to visualize, because the lattice points are the corners of a packing of space with unit cubes, the fact that there are rotation axes in so many directions actually means (as we shall see) that cubic lattices can pop up in unexpected places!

Experiment 2.4 This may be a rather difficult experiment to perform. Take the square lattice on Side 2 of the Lattice Card, call the red square overlay L_0, the blue square overlay L_1 and the green square overlay L_2. Support the overlays at the correct distances above the card to form a primitive cubic lattice. (Lumps of Blu-Tack or something similar will be very useful here.) Then align your eye along as many rotation axes as you can see. (You should be able to pick out axes going through opposite corners of cubes, and you should also be able to align sets of edges, one in each layer, and see that the axis through their midpoints is a rotation axis of order 2.) ♦

Case 5 $L(\mathbf{a}, \mathbf{b})$ is a hexagonal lattice

Finally, let $L(\mathbf{a}, \mathbf{b})$ be a plane hexagonal lattice. Choosing $\mathbf{c}$ to have zero offset and 'random' length gives us an analogous situation to the cases of the rectangular and square lattices — as you may have guessed, the rotational and reflection symmetries of $L(\mathbf{a}, \mathbf{b})$ correspond to rotation symmetries of $L(\mathbf{a}, \mathbf{b}, \mathbf{c})$ about the z-axis, and reflections through vertical planes, to give the point group

$$H = \Gamma(\mathrm{DIH}_6).$$

You may wish to perform Experiment 2.3 again, this time with the hexagonal lattice on Side 2 of the Lattice Card, and the corresponding overlays.

The lattice $L(\mathbf{a}, \mathbf{b}, \mathbf{c})$ is in this case called a **hexagonal lattice** (or a **hexagonal space lattice**, to distinguish it from a hexagonal plane lattice), and any crystal pattern with an associated lattice having point group $\Gamma(\mathrm{DIH}_6)$ belongs to the **hexagonal crystal system**.

2.3 Space lattices with non-zero offset

So far, we have encountered five crystal systems: monoclinic, orthorhombic, tetragonal, cubic and hexagonal. If we consider lattices $L(\mathbf{a}, \mathbf{b}, \mathbf{c})$ such that $\mathbf{c}$ has non-zero offset relative to $\{\mathbf{a}, \mathbf{b}\}$, we can obtain two more systems (i.e. two more possibilities for the point group).

Consider the effect of offsetting $L_1(\mathbf{a},\mathbf{b})$ by a random translation.

Experiment 2.5 Consider *any one* of the five plane lattices on the Lattice Card, and select the appropriate red and blue lattice overlays. If you place the red overlay (call this L_0) directly over the appropriate lattice of the Lattice Card, then place the blue overlay (call this L_1) over the first, offset a little way so that the lattice points of the blue overlay are all a short distance away from those of the red overlay, for example with offset $(\lambda, \mu) = \left(\frac{1}{6}, \frac{1}{7}\right)$, then you will find that the pair of overlays has no non-trivial rotation symmetry — no matter where you insert the pin, you cannot hit a point that is a rotation centre *both* for the red *and* for the blue overlay. Similarly, there is no reflection plane that sends the lattice points of L_0 to themselves *and also* sends those of L_1 to themselves. The inversion σ_O (which maps L_0 to itself, and L_1 to L_{-1}) is, however, still a symmetry. ♦

The inversion σ_O and the identity are now (in general) the *only* symmetries of $L(\mathbf{a},\mathbf{b},\mathbf{c})$ that fix the origin:

$$H = \{e, \sigma_O\}.$$

It *may* be possible to choose an offset and a vertical distance that happen to give a rotational symmetry *with the axis in a different direction* — but a *random* choice will not produce such symmetries.

The lattice $L(\mathbf{a},\mathbf{b},\mathbf{c})$ is in this case called a **triclinic lattice**, and any crystal system whose associated lattice has point group $\{e, \sigma_O\}$ belongs to the **triclinic crystal system**.

The name *triclinic* means 'three planes', and refers to the fact that the planes defining the sides of the basic parallelepiped, i.e. the planes containing the pairs $\{\mathbf{a},\mathbf{b}\}, \{\mathbf{b},\mathbf{c}\}$ and $\{\mathbf{a},\mathbf{c}\}$, are of equal significance in such a system.

Experiment 2.6 This one is more fun! Look at the hexagonal lattice on Side 2 of the Lattice Card. The large black dots on the card are the lattice points, while the smaller open dots are offset from the lattice points by $\frac{1}{3}(\mathbf{a}+\mathbf{b})$ and constitute one orbit of 3-centres. The small crosses are offset from the lattice points by $\frac{2}{3}(\mathbf{a}+\mathbf{b})$ and constitute the other orbit of 3-centres.

Place the red hexagonal lattice overlay over the card in such a way that its lattice points line up over those of the card, then place the blue hexagonal lattice overlay over the red one in such a way that its lattice points line up over the open dots on the card. Then place the green hexagonal lattice overlay over the others so that its lattice points line up over the crosses on the card. This corresponds to a space lattice based on stacking copies of the hexagonal plane lattice with an offset of $\left(\frac{1}{3}, \frac{1}{3}\right)$.

Insert a pin through the overlays and the card, piercing the card at any one of its lattice points, near the centre. Now rotate the overlays together. Note carefully how much rotation is required before the lattice points of the red overlay again lie over those of the card, while those of the blue overlay again lie over the open dots of the card and those of the green overlay lie again over the crosses of the card.

You should find that the rotation is $2\pi/3$, that is, one-third of a turn. (After one-sixth of a turn, the red overlay is correctly aligned again, but the other two overlays have exchanged the positions of their lattice points.) This corresponds to the fact that the origin of a hexagonal plane lattice is a 6-centre, while the point $\frac{1}{3}(\mathbf{a}+\mathbf{b})$ is a 3-centre. ♦

You saw this in Subsection 3.5 of *Unit GE3*

Thus, we have constructed a space lattice containing rotational symmetries of order 3 but no higher orders — a bit of a surprise, perhaps, since no plane lattice has 3 as its highest order of rotation.

Exercise 2.5

Is reflection in the xy-plane a symmetry of this lattice?

The answer to the above exercise tells us that H is not in this case dihedral; in fact

$$H = C_3 \cup \sigma_O C_3.$$

The lattice $L(\mathbf{a}, \mathbf{b}, \mathbf{c})$ is in this case called a **trigonal lattice**, and any crystal pattern with an associated lattice whose point group is $C_3 \cup \sigma_O C_3$ belongs to the **trigonal crystal system**.

Be careful, though! What about the vertical separation between the planes? The above analysis works satisfactorily if this is 'random' — but if you hold the blue overlay over the red overlay, offset as above, but at a certain critical height, then the vectors from a typical blue point, down to the three nearest red points, will be of equal length and at right angles to each other — try it!

Exercise 2.6

Assuming that $\mathbf{a} = (1, 0, 0)$ and $\mathbf{b} = \left(\frac{1}{2}, \frac{\sqrt{3}}{2}, 0\right)$, find the value of $\mathbf{c}$ (with $c_3 > 0$) such that $\mathbf{c} - \mathbf{0}$, $\mathbf{c} - \mathbf{a}$ and $\mathbf{c} - \mathbf{b}$ are orthogonal.

Using $\{\mathbf{c}, \mathbf{c} - \mathbf{a}, \mathbf{c} - \mathbf{b}\}$ as a basis for L, the result of Exercise 2.6 shows that L is in fact a primitive cubic lattice!

Experiment 2.7 To see how the actual cubes in this case are formed, four layers are necessary! Use the hexagonal lattice on the card as the bottom layer, then offset the red hexagonal overlay by $\left(\frac{1}{3}, \frac{1}{3}\right)$, the blue one by $\left(\frac{2}{3}, \frac{2}{3}\right)$ and the green one by $(1,1)$, which is the same as zero offset. Then any lattice point on the card is the bottom corner of a cube (to be thought of as balancing on that corner). Let us call this point O. The triangle of red lattice points that surround O represents the corners of the cube that are adjacent to the bottom corner. Then the oppositely oriented triangle of blue lattice points that also surround O represents three more corners of the cube, adjacent to the top corner, which is the green lattice point that lies directly over O (see Figure 2.6).

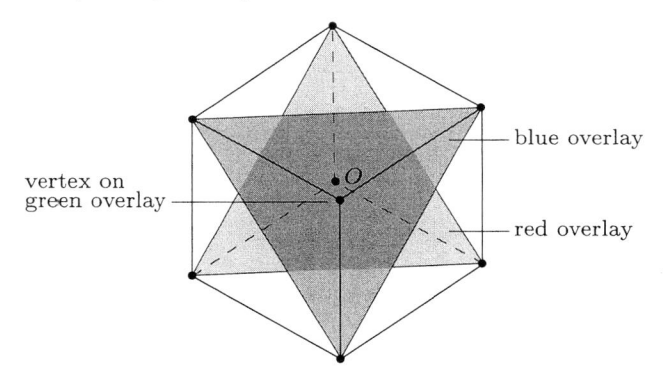

Figure 2.6 ♦

We have now found space lattices corresponding to all seven crystal systems. To summarize:

Name	Point group
monoclinic	$H = \{e, R_z[\pi], \sigma_O, Q_{xy}\} = C_2 \cup \sigma_O C_2$
orthorhombic	$H = \Gamma(\text{DIH}_2)$
tetragonal	$H = \Gamma(\text{DIH}_4)$
cubic	$H = \Gamma(\text{CUBE})$
hexagonal	$H = \Gamma(\text{DIH}_6)$
triclinic	$H = \{e, \sigma_O\} = C_1 \cup \sigma_O C_1$
trigonal	$H = C_3 \cup \sigma_O C_3$

However, we have actually seen eight different types of space lattice: we saw both a primitive and a base-centred orthorhombic lattice. In the next section, we shall see six more types of space lattice, all belonging to the monoclinic, orthorhombic, tetragonal or cubic systems.

3 THE BRAVAIS LATTICES

3.1 The base-centred monoclinic lattice

Let us return to the case of a plane parallelogram lattice $L(\mathbf{a}, \mathbf{b})$, and experiment with the offsets $\left(\frac{1}{2}, 0\right)$ and $\left(0, \frac{1}{2}\right)$.

Experiment 3.1 Look at the parallelogram lattice on Side 1 of the Lattice Card. Place the red parallelogram overlay (call this L_0) so that its lattice points lie directly over those of the card, then place the blue parallelogram overlay (call this L_1) over the first in such a way that its lattice points have offset $\left(\frac{1}{2}, 0\right)$; that is, they lie over the small open dots on the Lattice Card.

Now insert a pin through the overlays, piercing the card at a lattice point near the centre. Rotate the two overlays together, and note that the lattice points of each overlay are in their appropriate positions after a half-turn; this corresponds to the fact that the z-axis passes through a 2-centre both of L_0 and of L_1. In fact, the z-axis is an axis of rotation symmetry of order 2 for the whole lattice whose point group is therefore

Consider the origin to be the point you have chosen.

$$\{e, R_z[\pi], \sigma_O, Q_{xy}\} = C_2 \cup \sigma_O C_2,$$

the same as for the primitive monoclinic lattice. Therefore, this lattice also belongs to the monoclinic crystal system.

The even numbered layers (i.e. the plane lattices $L_0, L_2, L_4, \ldots$) form a space lattice whose offset is zero and whose basic parallelepipeds have vertical faces. Those faces that are parallel to the zx-plane have, at their centres, the lattice points of the odd numbered layers: $L_1, L_3, \ldots$ (see Figure 3.1).

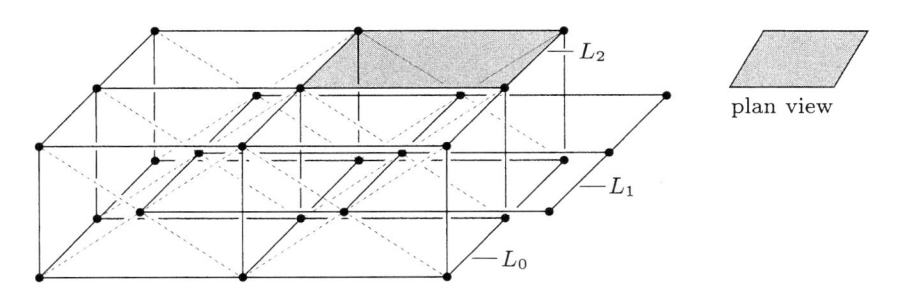

Figure 3.1 ♦

As with the base-centred orthorhombic lattice, we have in effect started with a primitive space lattice (in this case, primitive monoclinic) and placed a new lattice point at the centre of one face of each parallelepiped. Counting these faces as the base faces, we call this lattice a **base-centred monoclinic lattice**.

Experiment 3.2 Repeat Experiment 3.1, this time with offset $\left(0, \frac{1}{2}\right)$, corresponding to the crosses on the parallelogram lattice of the Lattice Card, and confirm that the space lattice that you obtain is again base-centred monoclinic. ♦

3.2 Body-centred space lattices

Suppose we try the offset $(\frac{1}{2}, 0)$ in conjunction with the rectangular lattice; what happens then? Figure 3.2 shows what we obtain.

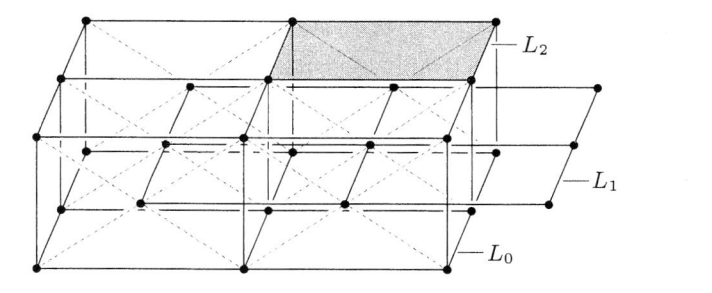

Figure 3.2

The even numbered layers now form a primitive orthorhombic lattice, and (just as in the base-centred monoclinic case) the vertical faces lying parallel to the zx-plane have the lattice points of the odd numbered layers at their centres.

Exercise 3.1

Identify the resulting lattice type as one that you have seen already.

The same conclusion would follow, of course, for the offset $(0, \frac{1}{2})$; but suppose that we now try the offset $(\frac{1}{2}, \frac{1}{2})$; what then? Once again, the even numbered layers have zero offset, and form a primitive orthorhombic lattice. This time, however, the lattice points of the odd numbered layers fall *at the centres* of the rectangular parallelepipeds formed by the even numbered layers (see Figure 3.3).

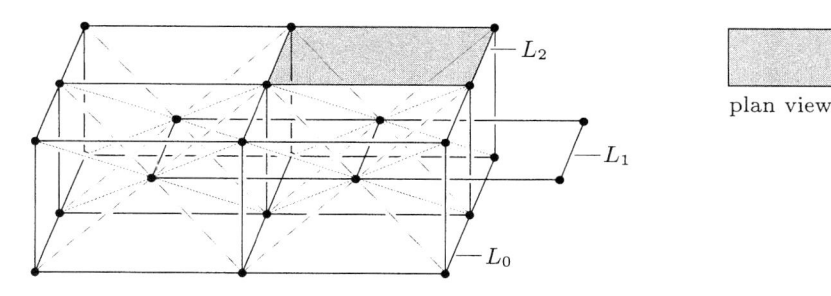

Figure 3.3

Experiment 3.3 You should be able to achieve this effect, using the three rectangular lattice overlays. Place the red and the green rectangular lattice overlays directly over the rectangular lattice on the card, with the blue overlay sandwiched between the other two and offset by $(\frac{1}{2}, \frac{1}{2})$ (that is, with the lattice points over the stars at the centres of the rectangles of the card). You should be able to see that the blue lattice points are at the centres of the boxes that would be formed if the overlays were evenly spaced vertically.

Now consider the symmetries of this lattice, in particular, the rotations and reflections that fix the origin. You should be able to repeat the rotations of the primitive and base-centred orthorhombic lattices, and be able to see that the reflection planes are also the same. ♦

Thus, although the full symmetry group is different from either of these, the point group is the same, so the crystal system is still orthorhombic.

This lattice is called a **body-centred orthorhombic lattice**.

Experiment 3.4 Consider the square lattice on Side 2 of the Card. Using the three square lattice overlays, repeat Experiment 3.3.

You should confirm that all the symmetries of the tetragonal system are still present.

Moreover, if $\mathbf{c} = \left(\frac{1}{2}, \frac{1}{2}, \frac{1}{2}\right)$, so that the vertical separation as well as the offset is $\frac{1}{2}$, then the even numbered layers form a cubic lattice, with the lattice points of the odd numbered layers at the centres of the cubes. ◆

It should be clear that the presence of these points at the centres of the parallelepipeds does not affect the point group. We have a **body-centred tetragonal lattice** in the case when the vertical separation is random, and a **body-centred cubic lattice** when the vertical separation is $\frac{1}{2}$ (see Figure 3.4).

body-centred tetragonal

body-centred cubic

plan view

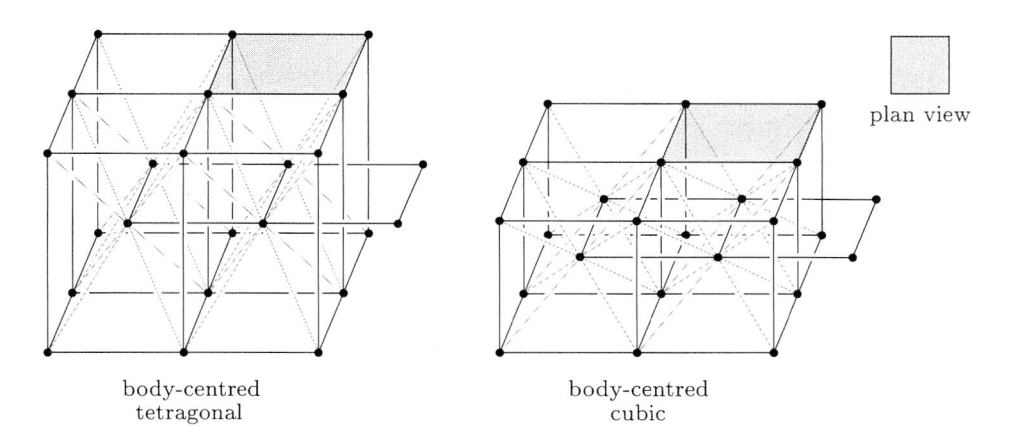

Figure 3.4

You have now seen twelve types of space lattice — seven primitive, two base-centred and three body-centred. Before we go on to consider the last two types, it is worth explaining why we do *not* get (for example) a base-centred tetragonal type, or a body-centred monoclinic type.

It is perfectly possible to do the constructions that 'ought' to provide these types, and we do indeed get perfectly good space lattices. The only problem is that they are not new types!

For example, if we take the square lattice $L(\mathbf{a}, \mathbf{b})$ with $\mathbf{a} = (1, 0)$ and $\mathbf{b} = (0, 1)$, and stack with an offset of $\left(\frac{1}{2}, 0\right)$ as if to try for a base-centred tetragonal lattice, then we do indeed obtain a system where the even numbered layers form a primitive tetragonal lattice and the odd numbered layers contribute lattice points to the centres of certain faces.

You may wish to use the card and overlays to try this out.

The only trouble is that they interfere with the tetragonal symmetry! Rotation through a quarter-turn about the z-axis is no longer a symmetry because it takes centred to non-centred faces. In fact, we are back with a base-centred orthorhombic lattice again.

What about trying to have a body-centred monoclinic lattice? The problem here is not a loss of symmetry, but simply the fact that we can realize it as a base-centred monoclinic lattice by a different choice of basis, as the next exercise shows.

Exercise 3.2

Let **a** and **b** be in the xy-plane, and let $\mathbf{c} = \frac{1}{2}(\mathbf{a}+\mathbf{b}) + (0,0,c)$, so that the offset is $\left(\frac{1}{2},\frac{1}{2}\right)$. Then the even numbered layers form a primitive monoclinic lattice, with the lattice points of the odd numbered layers at the centres of the parallelepipeds. Show that if the basis of the layers is changed to $\{\mathbf{a}, \mathbf{a}+\mathbf{b}\}$, then the parallelepipeds formed by the even numbered layers now have the lattice points of the odd numbered layers at the centres of certain faces.

You may wish to confirm experimentally the result of Exercise 3.2. The stars at the centres of the parallelograms on the lattice allow you to place an overlay at an offset of $\left(\frac{1}{2},\frac{1}{2}\right)$ relative to the original basis, and you can see that this is an offset of $\left(0,\frac{1}{2}\right)$ relative to the basis $\{\mathbf{a}, \mathbf{a}+\mathbf{b}\}$ if you draw a basic parallelogram of this basis on the card.

3.3 Face-centred space lattices

Our last two lattices are formed by taking a primitive lattice and adding extra lattice points to obtain a lattice point at the centre of *every* face of the parallelepipeds of the primitive system. This can always be done, but in fact there are only two primitive systems in which doing this gives a space lattice type that we have not seen before. These are the orthorhombic and cubic systems, shown in Figure 3.5.

face-centred orthorhombic   plan view   face-centred cubic   plan view

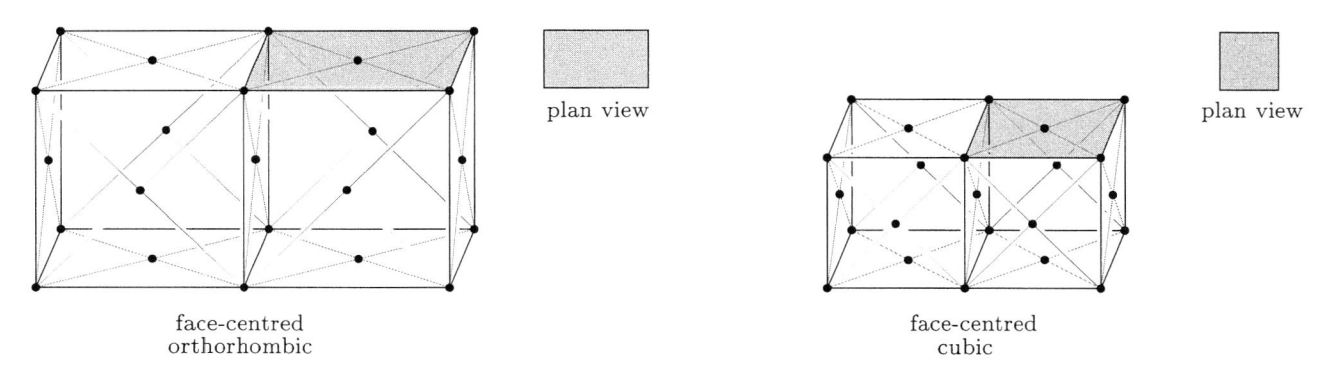

Figure 3.5

It should be fairly clear without further experimentation that these systems of points have the same point groups as the orthorhombic and cubic lattices respectively. What is not so obvious is that they are space lattices at all!

Suppose we start with the space lattice $L(\mathbf{a}, \mathbf{b}, \mathbf{c})$, then add further points at the centres of the faces of all the parallelepipeds. This means that there must be further lattice points at $\left(\left(n+\frac{1}{2}\right)\mathbf{a} + \left(m+\frac{1}{2}\right)\mathbf{b} + r\mathbf{c}\right)$, at $\left(\left(n+\frac{1}{2}\right)\mathbf{a} + m\mathbf{b} + \left(r+\frac{1}{2}\right)\mathbf{c}\right)$, and at $\left(n\mathbf{a} + \left(m+\frac{1}{2}\right)\mathbf{b} + \left(r+\frac{1}{2}\right)\mathbf{c}\right)$, for all triples n, m, r of integers. The question is whether there is a basis whose integer combinations give exactly these lattice points, and no others.

The answer is that the basis $\left\{\frac{1}{2}\mathbf{a}+\frac{1}{2}\mathbf{b}, \frac{1}{2}\mathbf{a}+\frac{1}{2}\mathbf{c}, \frac{1}{2}\mathbf{b}+\frac{1}{2}\mathbf{c}\right\}$ does the trick.

Exercise 3.3

Let us denote the elements of the above basis by $\mathbf{d}$, $\mathbf{e}$ and $\mathbf{f}$, respectively. Show that $\{\mathbf{d}, \mathbf{e}, \mathbf{f}\}$ generates the face-centred lattice as required.

This is an optional exercise. It is quite difficult; you have been warned!

These two space lattices are, of course, called the **face-centred orthorhombic lattice** and the **face-centred cubic lattice** respectively.

In particular, this means that the face-centred lattice derived from the cubic lattice $L((1,0,0),(0,1,0),(0,0,1))$ has a basis

$$\{\mathbf{a}, \mathbf{b}, \mathbf{c}\} = \left\{\left(\tfrac{1}{2}, \tfrac{1}{2}, 0\right), \left(\tfrac{1}{2}, 0, \tfrac{1}{2}\right), \left(0, \tfrac{1}{2}, \tfrac{1}{2}\right)\right\}.$$

The vectors $\mathbf{a}, \mathbf{b}, \mathbf{c}$ here correspond to the vectors $\mathbf{d}, \mathbf{e}, \mathbf{f}$ in Exercise 3.3.

Exercise 3.4

Show that the four points $\mathbf{0}, \mathbf{a}, \mathbf{b}, \mathbf{c}$ form the vertices of a regular tetrahedron.

The significance of this exercise is that if we orient the lattice so that $\mathbf{0}, \mathbf{a}$ and $\mathbf{b}$ lie in the xy-plane, then $L(\mathbf{a}, \mathbf{b})$ is a hexagonal plane lattice and $\mathbf{c}$ has offset $\left(\tfrac{1}{3}, \tfrac{1}{3}\right)$. We have seen this before! If the height of L_1 above the xy-plane is chosen at random, then (as we saw) we obtain a trigonal lattice, whereas if it is chosen so that $\mathbf{c}$, $\mathbf{c} - \mathbf{a}$ and $\mathbf{c} - \mathbf{b}$ are orthogonal, we obtain a primitive cubic lattice. By raising the height slightly, so that $\mathbf{0}, \mathbf{a}, \mathbf{b}$ and $\mathbf{c}$ form the vertices of a regular tetrahedron, we obtain a face-centred cubic lattice.

You will meet the face-centred cubic lattice again (and the hexagonal lattice) in the video programme for this unit.

Exercise 3.5

Assuming that $\mathbf{a}$ and $\mathbf{b}$ are unit vectors, how far above L_0 must L_1 be in order that $\mathbf{0}, \mathbf{a}, \mathbf{b}$ and $\mathbf{c}$ should form the vertices of a regular tetrahedron?

To summarize, the fourteen different types of space lattice are as follows.

Crystal system	Point group	Primitive	Base-centred	Body-centred	Face-centred
monoclinic	$\{e, R_z[\pi], \sigma_O, Q_{xy}\}$	✓	✓		
orthorhombic	$\Gamma(\mathrm{DIH}_2)$	✓	✓	✓	✓
tetragonal	$\Gamma(\mathrm{DIH}_4)$	✓		✓	
cubic	$\Gamma(\mathrm{CUBE})$	✓		✓	✓
hexagonal	$\Gamma(\mathrm{DIH}_6)$	✓			
triclinic	$\{e, \sigma_O\}$	✓			
trigonal	$C_3 \cup \sigma_O C_3$	✓			

We do not usually use the word 'primitive' for the hexagonal, triclinic or trigonal lattices, as there are no 'non-primitive' lattices in these systems from which they need to be distinguished.

However, it should be noted that some classification systems combine the hexagonal and the trigonal Bravais lattices into a single 'hexagonal crystal system', in which case the hexagonal lattice *is* referred to as *primitive hexagonal* and the trigonal lattice is then termed *rhombohedral hexagonal*.

4 POLYHEDRA

4.1 The semi-regular polyhedra

The concept of a *polyhedron* is simply the three-dimensional analogue of that of a polygon.

> **Definition 4.1 Polyhedron**
>
> A **polyhedron** is a solid whose surface consists of a finite number of faces, each a polygon.

For example, the regular solids and prisms which you studied in *Unit GE5* are polyhedra, but the dihedron is not, as it is a marked sphere (whose surface is curved).

Plato's name is associated with the five regular solids, and that of Archimedes with the *semi-regular polyhedra*, which are defined as follows.

> **Definition 4.2 Semi-regular polyhedra**
>
> The **semi-regular polyhedra**, or **Archimedean solids**, are polyhedra whose faces are regular polygons, not all of the same degree, but such that the degrees of the polygons meeting at each vertex are the same, in the same cyclic order.

The *degree* of a polygon is its number of sides.

Let us put this another way. Suppose we generalize the idea of *vertex type* which we met in *Unit IB1*, to apply to the faces meeting at a vertex of a polyhedron as well as to the tiles meeting at a vertex of a tiling. Then a polyhedron can be said to be *vertex-uniform* if all its vertices have the same type. We may now define a *semi-regular polyhedron* to be a polyhedron which is not one of the five regular solids, but which is vertex-uniform and all of whose faces are regular polygons.

An easy way to construct semi-regular polyhedra is to start with regular polyhedra and to chop off their vertices in a regular way.

If we chop the corners off a cube using cuts very close to the vertices, then each vertex is replaced by a small triangular face, and the square faces become octagons.

However, the octagons are not regular unless the cutting planes have been carefully chosen. Not only must each cutting plane be perpendicular to the line from the centre of the cube to the vertex being chopped, but their distances from the vertices must be selected so that the new edges of each octagon are the same length as what is left of the old edges. It is perfectly possible to do this; the result, known as a **truncated cube**, is a semi-regular polyhedron whose faces are regular octagons and equilateral triangles.

We do not remove any of the symmetry of the original cube. Every cutting plane is perpendicular to an axis of rotational symmetry, so the axis in question continues to be an axis of rotation, but it now passes through the centres of a pair of opposite (triangular) faces. Thus, the symmetry group of the truncated cube is just $\Gamma(\text{CUBE})$.

What if we now move our cutting planes nearer to the centre of the original cube? The edges left over from those of the original cube become progressively shorter, until they disappear altogether and the octagons become squares.

The stages of truncation are illustrated in Figure 4.1.

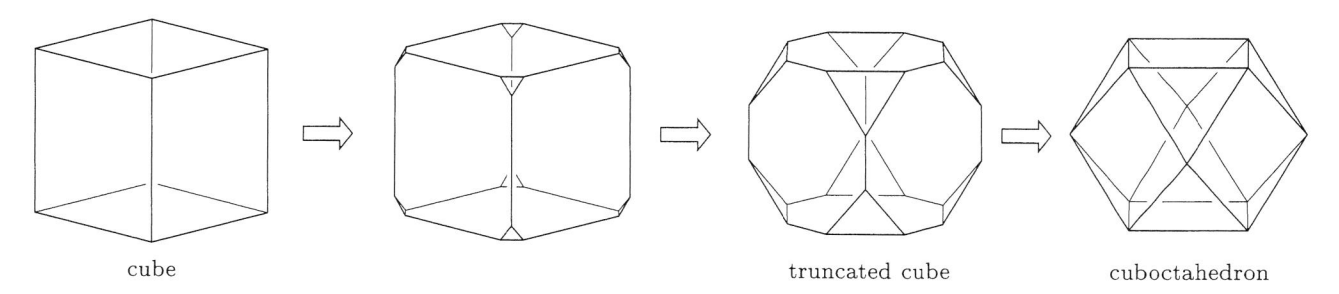

Figure 4.1

The last solid depicted in Figure 4.1 is a new semi-regular polyhedron, called a **cuboctahedron**. (As you will see, it can also be obtained by truncating an octahedron, so in a sense it is midway between a cube and an octahedron — hence the name.)

Exercise 4.1

(a) Write down the symmetry group of the cuboctahedron.
(b) Write down the vertex types of the cube, the truncated cube and the cuboctahedron.

What about truncating an octahedron? The vertices of an octahedron are of degree 4, and this time the new faces resulting from chopping away the vertices are squares. The original faces double their degree and become hexagons; chopping deeply enough makes the hexagons regular. Thus we obtain a semi-regular polyhedron, called a **truncated octahedron**, whose faces are squares and regular hexagons.

If we truncate more deeply, so that the original edges of the octahedron disappear completely, then we obtain a polyhedron whose faces are squares and equilateral triangles — in fact, this is the cuboctahedron again.

The stages of truncation are illustrated in Figure 4.2.

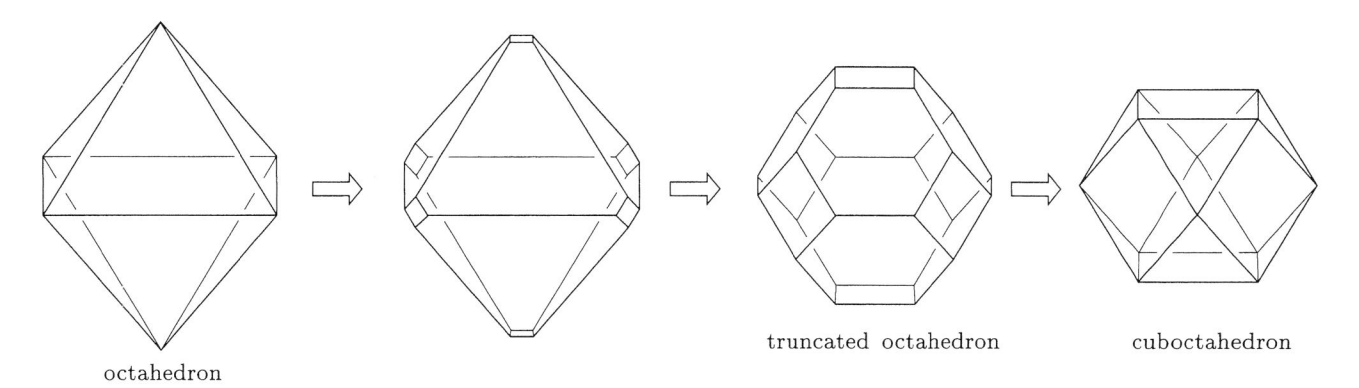

Figure 4.2

Exercise 4.2

(a) What is the symmetry group of the truncated octahedron?
(b) What is the vertex type of the truncated octahedron?

Next, let us start with the tetrahedron. Again, we can chop off each vertex in such a way as to leave portions of the original edges equal in length to the edges of the new triangles. Thus, the result (called a **truncated tetrahedron**) is a semi-regular polyhedron whose faces are equilateral triangles and regular hexagons. By a similar argument to that for the truncated cube, the symmetry group is $\Gamma(\text{TET})$.

Once again, if we truncate more deeply, so that the portions of the original edges of the tetrahedron disappear altogether, we can obtain another polyhedron. This time, though, the original faces are cut down to equilateral triangles and the new faces are also triangles, so instead of obtaining a semi-regular polyhedron, we obtain one of the regular solids — namely the octahedron!

The stages of truncation are illustrated in Figure 4.3.

tetrahedron   truncated tetrahedron   octahedron

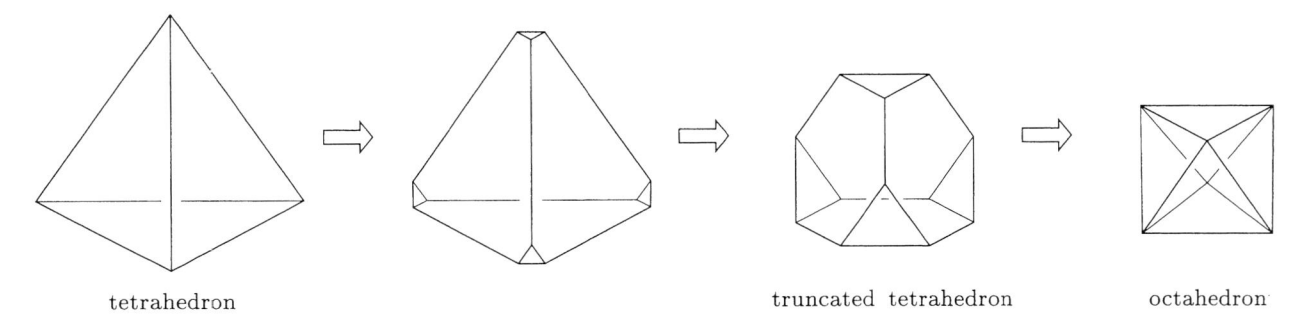

Figure 4.3

How is this consistent with our claim that truncation retains all the symmetry of the original? The answer is that, in this particular case, we do indeed retain all the symmetry of the original tetrahedron, but we also add *further* symmetries.

Since the symmetry group of the octahedron is the same as that of the cube, this is rather a roundabout way of showing that $\Gamma(\text{TET}) \subset \Gamma(\text{CUBE})$ — but it works!

Exercise 4.3

Find the vertex type of the truncated tetrahedron.

The other two regular solids are the dodecahedron and the icosahedron. Here too, we can truncate each of these to various depths. The dodecahedron has faces of degree 5 and vertices of degree 3, so we can truncate it in a way that produces regular faces of degrees 10 and 3 — the **truncated dodecahedron**. In a similar way, we can obtain the **truncated icosahedron**.

As with the dual pair consisting of the cube and the octahedron, the result of a deeper truncation of either the dodecahedron or the icosahedron is to produce another semi-regular polyhedron that is in a sense midway between the two, called the **icosidodecahedron**.

The stages of truncation are shown in Figures 4.4 and 4.5.

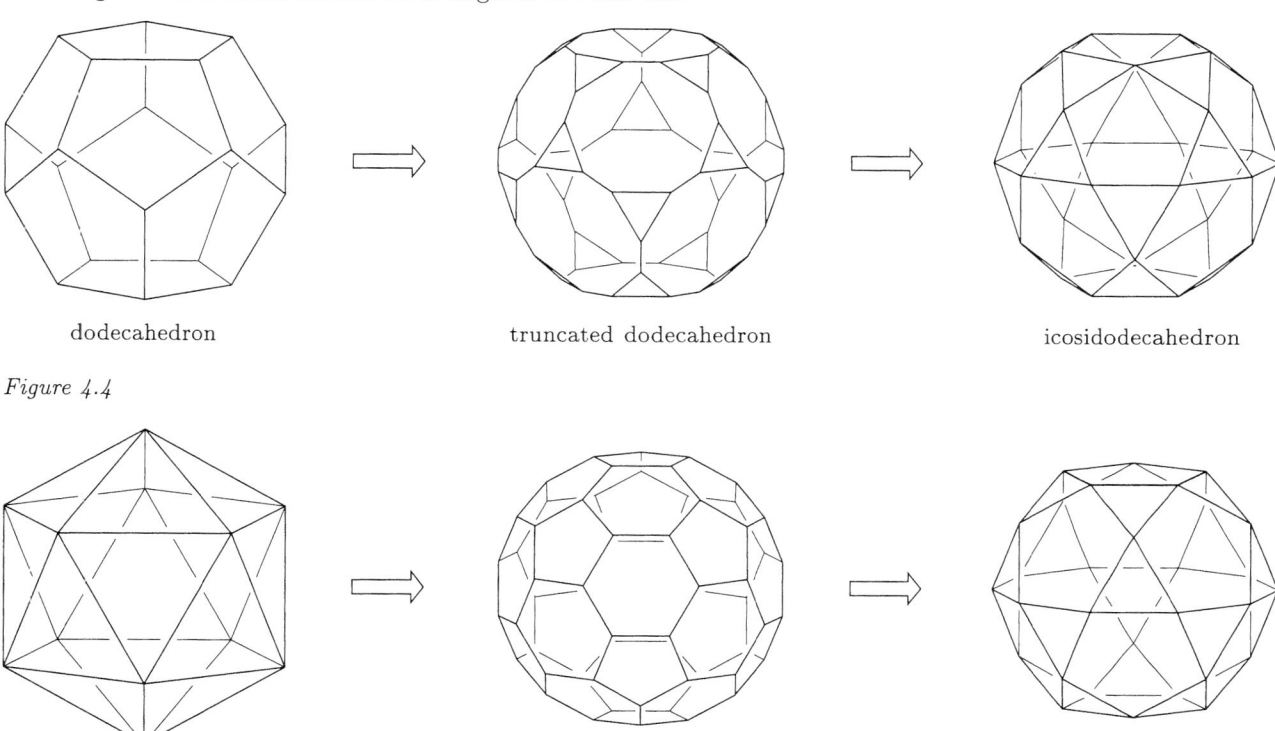

dodecahedron ⇒ truncated dodecahedron ⇒ icosidodecahedron

Figure 4.4

icosahedron ⇒ truncated icosahedron ⇒ icosidodecahedron

Figure 4.5

Exercise 4.4

Write down the vertex types of the truncated dodecahedron, the truncated icosahedron and the icosidodecahedron. What is their symmetry group?

There are six further semi-regular polyhedra that can be obtained by more complicated truncations of the regular solids; these are shown in Figure 4.6.

snub cube
(snub cuboctahedron)

small
rhombicosidodecahedron

snub dodecahedron
(snub icosidodecahedron)

great rhombicuboctahedron
(truncated cuboctahedron)

small
rhombicuboctahedron

great rhombicosidodecahedron
(truncated icosidodecahedron)

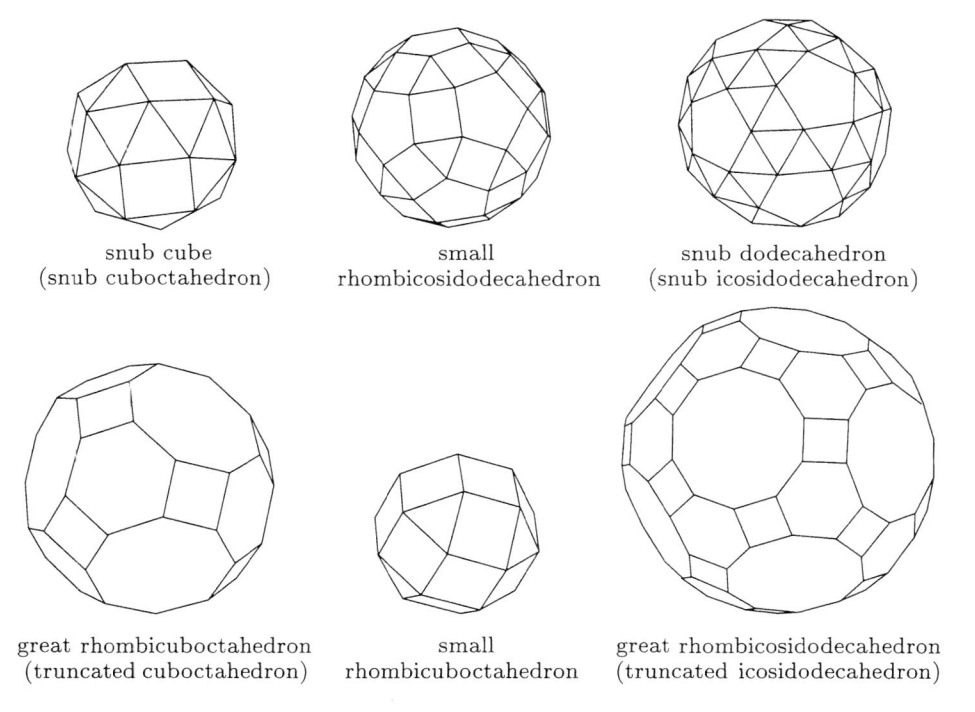

Figure 4.6

In addition, there are the **prisms** and **antiprisms**. You have seen prisms in
Unit GE5; for each $n > 2$ the n-prism has two parallel faces consisting of
congruent regular n-gons and a 'belt' of n rectangles. If the original n-gons
are the correct distance apart, the rectangles are squares and we have a
semi-regular polyhedron. The semi-regular 6-prism is shown in Figure 4.7.

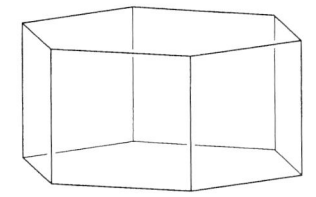

Figure 4.7

The antiprisms are somewhat similar to the prisms. Two parallel congruent
regular n-gons are taken; but this time, one is rotated by π/n with respect
to the other and they are connected by a 'belt' of $2n$ isosceles triangles. If
the original n-gons are the correct distance apart, the triangles are
equilateral and we have a semi-regular polyhedron. The semi-regular
5-antiprism is shown in Figure 4.8.

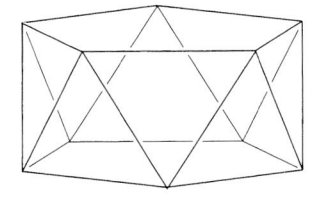

Figure 4.8

Exercise 4.5

One of the regular solids is a prism and one is an antiprism. Identify these.

Finally, all the regular and semi-regular polyhedra have **dual polyhedra**,
obtained from the originals by placing a vertex of each dual at the centre of
each face of the original, in much the same way that dual tilings are
constructed.

For the regular solids, the duality construction is graphically illustrated in
the video programme VC4A, which you should have viewed in conjunction
with *Unit GE5*. We shall not study these duals in general, but there is one
particular dual polyhedron worth mentioning. This is the dual of the
cuboctahedron, and is shown in Figure 4.9.

These duals should really be called
face duals, to distinguish them
from the *space dual* of a filling of
space with polyhedra. As you will
see in the video programme VC4B,
space duals are constructed by
placing vertices at the centres of
the polyhedra themselves, rather
than at the centres of their faces.

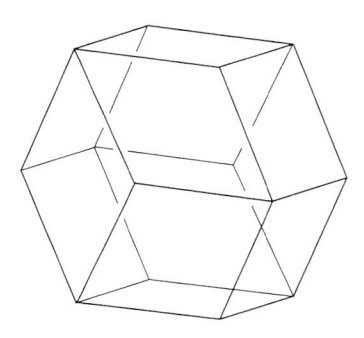

Figure 4.9

Its faces are twelve rhombuses, and it is called the **rhombic dodecahedron**, or sometimes **rhombidodecahedron**. There is no evidence that it was known to the ancient Greeks, but (as the video programme shows) it is used by the bees!

The three-dimensional analogues of polygonal tilings are ways of filling space with polyhedra. Although the general problem of classifying these is much more difficult than even the corresponding classification problem for tilings, a certain amount is known about filling space with regular and semi-regular polyhedra. This is the subject of the final video programme of the course.

4.2 Space-filling polyhedra (video subsection)

You should now watch the video for this unit, VC4B.

Please turn now to the Video Notes, do the pre-video work indicated there, watch the video, and then do the post-video work before continuing your study of this unit.

4.3 Counting with groups revisited

In *Unit GE1*, you studied a powerful technique for counting the number of essentially different colourings (or, more generally, labellings) of a geometric configuration. However, we did not have all that many interesting symmetry groups at our disposal at that point in the course. In particular, the analysis of the colourings of the faces of a cube was right at the limits of our analysis, and we presented it in an optional appendix.

Now that you have studied *Unit GE5* and most of *Unit GE6*, though, you should be able to take this kind of thing in your stride! How about the following?

Example 4.1

How many essentially different ways are there of colouring the faces of a truncated cube using the colours black and white?

Look at Figure 4.1.

Assuming that 'essentially different' means 'in the same orbit under the direct symmetry group', we may apply the analysis given in the video programm VC4A, as the group is still the rotation group of the cube. This time, however, we must consider permutations of *all* the faces of the truncated cube. But each triangular face of the truncated cube just corresponds to a vertex of the original cube, while each octagonal face corresponds to a (square) face of the original cube.

Thus, a rotation of order 4 about an axis through the centres of opposite faces of the original cube gives (for the cube) a 4-cycle and two 1-cycles for the faces and two 4-cycles for the vertices. Therefore, for the *truncated* cube, the permutation of the faces consists of three 4-cycles and two 1-cycles. Thus (before division by the order of the group) the contribution of each such symmetry to the cycle index is $x_1^2 x_4^3$. ◊

Exercise 4.6

Find the contribution for a symmetry of each of the following types:

(a) rotation of order 2 about an axis through opposite faces of the original cube;

(b) rotation of order 2 about an axis through the midpoints of opposite sides of the original cube;

(c) rotation of order 3 about an axis through opposite vertices of the original cube.

Example 4.1 continued

We know from the video programme that there are six group elements of the first type, three of the second, six of the third type and eight of the fourth. Finally, there is the identity, which contributes x_1^{14}. Thus, the cycle index is

$$P_G(x_1, x_2, x_3, x_4) = \tfrac{1}{24}\left(x_1^{14} + 6x_1^2 x_4^3 + 3x_1^2 x_2^6 + 6x_2^7 + 8x_1^2 x_3^4\right).$$

On substituting 2 for each variable x_i, we obtain the required number of colourings as 776. ♦

Exercise 4.7

A set of six rods of equal length is made into a tetrahedral framework. How many equivalence classes of colourings are there, using three colours?

Exercise 4.8

How many equivalence classes of colourings are there of the faces of a truncated tetrahedron, using m colours?

Example 4.2

Each face of a cube is divided into two triangles by a diagonal, in such a way that three diagonals meet at each one of a pair of opposite vertices as shown in Figure 4.10.

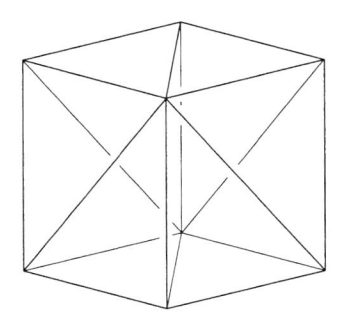

Figure 4.10

How many equivalence classes of colourings are there of the triangles, using two colours?

This time, the symmetries are considerably reduced. Any rotational symmetry must map the two vertices at which triples of diagonals meet, either to themselves or to each other. (Let us call these the 'special' vertices.) Thus the direct symmetry group is $\Gamma^+(DIH_3)$.

There are twelve triangles, so the identity contributes x_1^{12} to the cycle index (before division by the group order). Each rotation of order 3, keeping the 'special' vertices fixed, contributes x_3^4, while each of the rotations that map the 'special' vertices to each other is a rotation of order 2 about an axis through opposite edges, and thus produces six 2-cycles on the triangles, giving x_2^6. Thus,

$$P_G(x_1, x_2, x_3) = \tfrac{1}{6}\left(x_1^{12} + 2x_3^4 + 3x_2^6\right),$$

giving the number of equivalence classes of colourings as 720. ♦

Exercise 4.9

The faces of a cube are each divided in half, in such a way that four of the dividing lines are parallel. The other two dividing lines are parallel to each other, as shown in Figure 4.11.

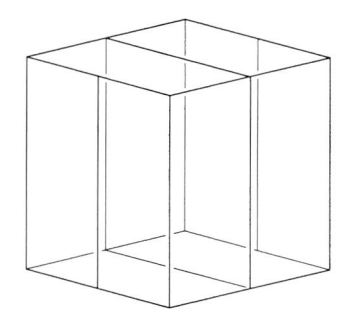

Figure 4.11

How many equivalence classes of colourings are there of the twelve rectangles so formed, using two colours?

5 CONCLUSION

5.1 Looking back

In the Geometry stream of this course, we have studied a number of types of geometric object in $\mathbb{R}^2$ and $\mathbb{R}^3$. In the main, we have concentrated on classifying the types of such objects, using their symmetry groups and other geometric properties. To summarize, we found:

11 types of Archimedean tiling, classifying by vertex type;

11 types of Laves tiling, classifying by tile type;

7 types of frieze pattern, classifying by the geometric properties of the symmetry group (although there are only four *isomorphism* classes of frieze groups);

81 types of transitive tiling, counting two as different if they have *either* different tile types *or* different symmetry group actions;

5 types of plane lattice, classifying by symmetry group action (but only four types if we classify by point group);

17 types of wallpaper pattern, classifying by symmetry group (and this time we do not have the ambiguity we had with the frieze groups — two wallpaper groups that differ in their geometric properties are actually non-isomorphic);

5 types of regular solid, classifying either by shape or by how the vertices, edges and faces fit together (but with only three different symmetry groups);

14 types of space lattice, classifying by symmetry group action (but only seven types if we classify by point group);

13 types of semi-regular polyhedron plus two infinite classes (the regular prisms and antiprisms), classifying either by shape or by how the vertices, edges and faces fit together.

We have also (in *Unit GE1*) seen how to count the essentially different colourings of a geometric configuration, and in the previous subsection of this unit we have studied some three-dimensional examples.

5.2 Looking forward

There are numerous other types of geometric structure which can be classified in ways similar to the above. Branko Grünbaum and Geoffrey Shephard have applied incidence symbols to the problem of classifying *vertex-transitive tilings* and *edge-transitive tilings*, and have also classified numerous other well-defined types of tilings and patterns in their book.

See the bibliography in the *Course Guide*.

What about three and more dimensions? We mentioned briefly that there are 230 distinct *crystallographic groups* — these are the three-dimensional analogues of the wallpaper groups. In dimensions higher still, little is known. It is known that the number of n-dimensional 'crystallographic groups' is finite for any finite n (by no means a trivial result, if you stop to think that, even in two dimensions, there are infinitely many distinct groups that fix a point, namely the groups C_n and D_n for each positive integer n). Moreover, in the case of four dimensions the number was recently discovered to be 4496 — but a computer had to be used. The numbers of distinct crystallographic groups in dimensions greater than four are not known.

What about the analogues of the frieze groups? Recall that a frieze group is not simply a one-dimensional analogue of a wallpaper group, as it is a group of *two*-dimensional isometries, forming the symmetry group of a *two*-dimensional structure. The 'one-dimensionality' arises from the fact that the translation group is generated by just one element, so that there is translational symmetry only in one direction. Thus, for any two positive integers m and n with $m < n$, we can consider the groups of n-dimensional isometries whose translation groups are generated by just m elements.

For $n = 3$, these groups are of some practical interest.

A structure such as a knitting pattern needs three-dimensional space for its existence (the threads must cross over each other) and, unlike a wallpaper pattern, turning a piece of fabric over will sometimes reveal a different pattern on the other side. The number of types of group with $m = 2$ and $n = 3$ is known — it is in fact 80.

What about $m = 1$ and $n = 3$? If we paint a pattern on a column of a building, for example, then (imagining the column to extend indefinitely far) we can obtain a pattern with translational repetition in only one direction — that of the column — but we can also organize for the pattern to repeat itself any number of times as we go round the column. (Think of a fluted column, for example; clearly there is no theoretical limit to the number of flutings that one can have, see Figure 5.1.) Thus, the number of possible groups here is infinite!

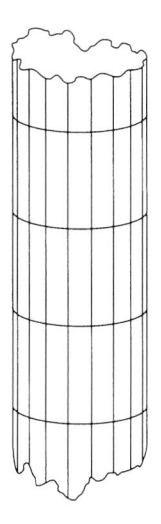

Figure 5.1

What about the analogues of tiling types? These are space-filling polyhedra. If we ask that they should all be congruent regular solids, then the cube is the only possibility; but if more than one type of regular solid is allowed, then (as we saw in the video programme) various possibilities arise. The analogue of the Archimedean tilings would be space fillings by more than one type of regular solid, but allowing only one vertex type and with each face of any one solid packed flush with a face of another. The only three possibilities here are the usual filling with cubes and the two types of 'octet' fillings which you saw in the video programme. If we allow our solids to be *either* regular *or* semi-regular, the number of possibilities rises to 23.

5.3 Revision exercises

The following exercises are intended to give you some practice with the type of question that you may be asked in Part 1 of the examination. They should help you to revise *Units IB1, IB3* and *GE1–GE4*.

Exercise 5.1

(a) Express in the form $t[\mathbf{p}]\, r[\theta]$ or $t[\mathbf{p}]\, q[\theta]$ the isometry

$$r[\pi]\, q[(1,-1),(1,1),3\pi/4].$$

(b) What type of isometry is this?

Exercise 5.2

The frieze shown in Figure 5.2 is of Type 7. Give in standard form the symmetries which map the motif labelled 1 to each of the other two labelled motifs. (Take the vertical line drawn through the frieze as the y-axis.)

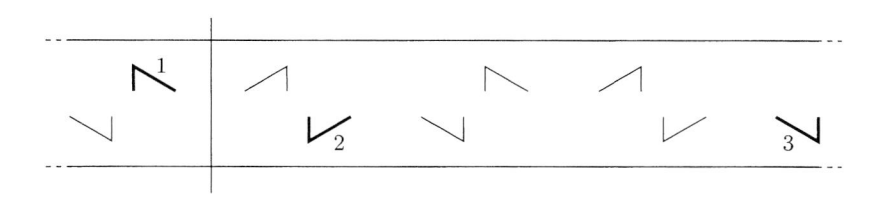

Figure 5.2

Exercise 5.3

(a) Write down the cycle index of the group $G = D_5$ acting on the edges of a regular pentagon.

(b) Write out the corresponding pattern inventory, supposing that two colours (labelled R and G) are available. You need *not* expand the inventory.

Exercise 5.4

(a) Write down the incidence symbol of the marked tiling $\mathcal{T}$, shown in Figure 5.3.

Figure 5.3

(b) Write down $n_t(\mathcal{T})$.

(c) Using a theorem from *Unit GE2*, or otherwise, find $n_v(\mathcal{T})$.

Exercise 5.5

The plane lattice shown in Figure 5.4 has basis $\{(2,0),(1,\sqrt{3})\}$.

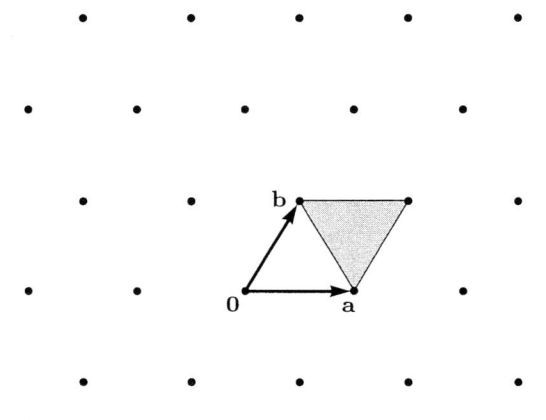

Figure 5.4

(a) If G_1 is the stabilizer of the origin and G_2 the stabilizer of the shaded triangle, write down (using any convenient notation) $G_1 \cap G_2$.

(b) Write down (using any convenient notation) a rotation and a glide reflection, each of which map the shaded triangle to the triangle whose vertices have vectors $\mathbf{0}, -\mathbf{a}, \mathbf{b} - \mathbf{a}$.

Exercise 5.6

(a) How many types of wallpaper pattern have rotations of order 2 but of no higher orders?

(b) Explain carefully how the theorem on rectangular and rhombic axes shows that this is the correct number.

SOLUTIONS TO THE EXERCISES

Solution 1.1

If the origin is placed at the bottom corner of a dark cube, then for the layer of cubes that sits on the xy-plane, the corresponding corners of dark cubes will be at $(0,0,0), (2,0,0), (4,0,0), \ldots$, then at $(1,1,0), (3,1,0), (5,1,0), \ldots$. Thus the lattice points associated with the dark cubes in this layer are the integer combinations of $(2,0,0)$ and $(1,1,0)$. The next layer up contains a dark cube with its bottom corner at $(0,1,1)$. Therefore the associated lattice can be generated by $\{(2,0,0), (1,1,0), (0,1,1)\}$.

This not the only solution; another is $\{(1,1,0), (1,-1,0), (0,1,1)\}$ and yet another is $\{(1,1,0), (0,1,1), (1,0,1)\}$.

Solution 1.2

The closest packing occurs when the centres of the discs lie on the vertices of a hexagonal plane lattice, the length of the basis vectors being equal to the diameter of the discs, as in the figure.

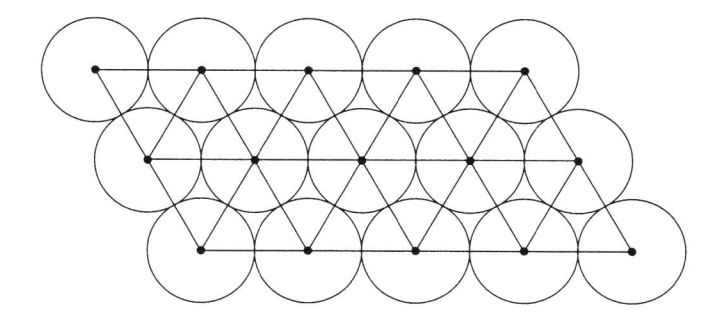

Solution 1.3

(a) $\begin{vmatrix} 1 & 0 & 0 \\ 1 & 1 & 0 \\ 1 & 2 & 3 \end{vmatrix} = 1 \times \begin{vmatrix} 1 & 0 \\ 2 & 3 \end{vmatrix} = 3.$

Thus the volume is 3.

(b) $\begin{vmatrix} 1 & 0 & 0 \\ 1 & 1 & 1 \\ 1 & 2 & 3 \end{vmatrix} = 1 \times \begin{vmatrix} 1 & 1 \\ 2 & 3 \end{vmatrix} = 1.$

Thus the volume is 1.

(c) $\begin{vmatrix} 1 & 1 & 1 \\ \frac{1}{2} & 1 & 2 \\ 1 & 3 & 9 \end{vmatrix} = 1 \times \begin{vmatrix} 1 & 2 \\ 3 & 9 \end{vmatrix} - 1 \times \begin{vmatrix} \frac{1}{2} & 2 \\ 1 & 9 \end{vmatrix} + 1 \times \begin{vmatrix} \frac{1}{2} & 1 \\ 1 & 3 \end{vmatrix}$
$= 3 - \frac{5}{2} + \frac{1}{2} = 1.$

Thus the volume is 1.

(d) $\begin{vmatrix} 1 & 1 & 0 \\ 1 & -1 & 0 \\ 1 & 2 & 3 \end{vmatrix} = 1 \times \begin{vmatrix} -1 & 0 \\ 2 & 3 \end{vmatrix} - 1 \times \begin{vmatrix} 1 & 0 \\ 1 & 3 \end{vmatrix}$
$= -3 - 3 = -6.$

Thus the volume is 6.

Solution 1.4

Only set (b) is a basis for L. (The volumes of the parallelepipeds defined by sets (a) and (d) are greater than 1, while set (c) does not consist of three lattice points — in particular, $\left(\frac{1}{2}, 1, 2\right)$ is not a point of L.)

Solution 1.5

(a) 0

(b) 0, as the offset may be chosen modulo 1.

(c) $\frac{1}{2}$

(d) $\frac{9}{10}$

(e) $\frac{2}{3}$, as $\frac{14}{3} = 4 + \frac{2}{3}$.

Solution 1.6

(a) Choosing $\mathbf{p} = (0, 1)$,
$$\mathbf{b} = \mathbf{p} + \tfrac{1}{2}(2, 0),$$
so $\omega = \tfrac{1}{2}$.

(b) This time, choosing $\mathbf{p} = \left(\tfrac{1}{2}, 0\right)$,
$$\mathbf{b} = \mathbf{p} + \tfrac{1}{3}\mathbf{a},$$
so $\omega = \tfrac{1}{3}$.

(c) Here we choose $\mathbf{p} = (-1, 1)$; then
$$\mathbf{b} = \mathbf{p} + \mathbf{a}.$$
Thus, $n + \omega = 1$, giving $n = 1$ and $\omega = 0$.

(d) Now we take $\mathbf{p} = \left(-\tfrac{3}{2}, \tfrac{3}{2}\right)$, giving
$$\mathbf{b} = \mathbf{p} + \tfrac{3}{2}\mathbf{a},$$
so that $n = 1$ and $\omega = \tfrac{1}{2}$.

Solution 1.7

(a) If L is square or rectangular, then it has a basis $\{\mathbf{a}, \mathbf{b}\}$ in which $\mathbf{b}$ is orthogonal to $\mathbf{a}$; thus, the offset of $\mathbf{b}$ is zero.

Conversely, if the offset of $\mathbf{b}$ is zero, then
$$\mathbf{b} = \mathbf{p} + (n + 0)\mathbf{a}$$
$$= \mathbf{p} + n\mathbf{a},$$
and since the transition matrix from $\{\mathbf{a}, \mathbf{p}\}$ to $\{\mathbf{a}, \mathbf{b}\}$, namely
$$\begin{bmatrix} 1 & n \\ 0 & 1 \end{bmatrix},$$
has determinant 1, $\{\mathbf{a}, \mathbf{p}\}$ is also a basis of L, with $\mathbf{p}$ orthogonal to $\mathbf{a}$. Thus, L is square or rectangular.

(b) The transition matrix from $\{\mathbf{b}, \mathbf{b} - \mathbf{a}\}$ to $\{\mathbf{a}, \mathbf{b}\}$ is
$$\begin{bmatrix} 0 & -1 \\ 1 & 1 \end{bmatrix},$$
which has determinant 1; thus $\{\mathbf{b}, \mathbf{b} - \mathbf{a}\}$ is indeed a basis of L. Now,
$$\mathbf{b} \cdot \mathbf{b} = \left(\mathbf{p} + \tfrac{1}{2}\mathbf{a}\right) \cdot \left(\mathbf{p} + \tfrac{1}{2}\mathbf{a}\right)$$
$$= \mathbf{p} \cdot \mathbf{p} + \tfrac{1}{2}\mathbf{a} \cdot \mathbf{p} + \mathbf{p} \cdot \tfrac{1}{2}\mathbf{a} + \tfrac{1}{4}\mathbf{a} \cdot \mathbf{a}$$
$$= \mathbf{p} \cdot \mathbf{p} + \tfrac{1}{4}\mathbf{a} \cdot \mathbf{a} \quad \text{(since } \mathbf{a} \text{ is orthogonal to } \mathbf{p}\text{)},$$
while
$$(\mathbf{b} - \mathbf{a}) \cdot (\mathbf{b} - \mathbf{a}) = \left(\mathbf{p} - \tfrac{1}{2}\mathbf{a}\right) \cdot \left(\mathbf{p} - \tfrac{1}{2}\mathbf{a}\right)$$
$$= \mathbf{p} \cdot \mathbf{p} + \tfrac{1}{4}\mathbf{a} \cdot \mathbf{a} \quad \text{(as before)}$$
$$= \mathbf{b} \cdot \mathbf{b}.$$

Thus, $||\mathbf{b}|| = ||\mathbf{b} - \mathbf{a}||$, which shows that L is rhombic, square or hexagonal.

Solution 1.8

We have

$$\begin{aligned}\mathbf{c} &= (0,0,2) + (2,3,0) \\ &= (0,0,2) + \tfrac{1}{3}\mathbf{a} + \tfrac{5}{3}\mathbf{b} \\ &= (0,0,2) + \tfrac{1}{3}\mathbf{a} + \left(1 + \tfrac{2}{3}\right)\mathbf{b},\end{aligned}$$

so the offset is $\left(\tfrac{1}{3}, \tfrac{2}{3}\right)$.

Solution 2.1

The rectangular and rhombic lattices — the group fixing the origin is D_2 for each of these.

Solution 2.2

The rotation $R_z[\pi]$ reverses the signs of the x- and y-coordinates. The central inversion σ_O reverses the signs of all three coordinates. Therefore the composite reverses the sign only of the z-coordinate. Geometrically, it is reflection in the xy-plane.

Solution 2.3

$$H^+ = \{e, R_x[\pi], R_y[\pi], R_z[\pi]\}.$$

Solution 2.4

The point group in this case is $\Gamma(\text{CUBE})$. There are three axes passing through centres of opposite faces, six through midpoints of opposite sides and four through opposite vertices, making a total of thirteen directions of axes of rotational symmetry.

For every one of these directions, there is an axis of rotational symmetry through the origin in that direction.

Solution 2.5

No. The layer L_1 is offset from the layer L_0 by $\left(\tfrac{1}{3}, \tfrac{1}{3}\right)$, but L_{-1} is offset from L_0 by $\left(\tfrac{2}{3}, \tfrac{2}{3}\right)$ (since $-\tfrac{1}{3} = (-1) + \tfrac{2}{3}$). Thus reflection in the xy-plane does not map the points of L_1 to points of L_{-1} (or to points of any other layer), and so is not a symmetry of the lattice.

Solution 2.6

As the offset is $\left(\tfrac{1}{3}, \tfrac{1}{3}\right)$, we know that

$$\begin{aligned}\mathbf{c} - \mathbf{0} &= \mathbf{c} \\ &= \tfrac{1}{3}\mathbf{a} + \tfrac{1}{3}\mathbf{b} + (0,0,c_3) \\ &= \left(\tfrac{1}{2}, \tfrac{\sqrt{3}}{6}, c_3\right),\end{aligned}$$

so that

$$\begin{aligned}\mathbf{c} - \mathbf{a} &= \left(-\tfrac{1}{2}, \tfrac{\sqrt{3}}{6}, c_3\right), \\ \mathbf{c} - \mathbf{b} &= \left(0, -\tfrac{\sqrt{3}}{3}, c_3\right).\end{aligned}$$

These three must be orthogonal, so, in particular,

$$\mathbf{c} \cdot (\mathbf{c} - \mathbf{a}) = 0,$$

giving

$$-\tfrac{1}{4} + \tfrac{3}{36} + c_3^2 = 0,$$

and the positive value for c_3 is therefore $\tfrac{1}{\sqrt{6}}$.

Solution 3.1

This is a base-centred orthorhombic lattice.

Solution 3.2

Let $\mathbf{b}' = \mathbf{a} + \mathbf{b}$. Then

$$\mathbf{c} = \tfrac{1}{2}\mathbf{b}' - (0,0,c),$$

and so has offset $\left(0, \tfrac{1}{2}\right)$ relative to the basis $\{\mathbf{a}, \mathbf{b}'\}$ of L_0. We have already seen that this gives a base-centred monoclinic lattice.

Solution 3.3

We first check that the original lattice points of $L(\mathbf{a}, \mathbf{b}, \mathbf{c})$ can be generated as integer combinations of the new basis. Now,

$$\mathbf{a} = \mathbf{d} + \mathbf{e} - \mathbf{f},$$
$$\mathbf{b} = \mathbf{d} + \mathbf{f} - \mathbf{e},$$
$$\mathbf{c} = \mathbf{e} + \mathbf{f} - \mathbf{d}.$$

Thus any integer combination of $\mathbf{a}$, $\mathbf{b}$ and $\mathbf{c}$ can also be obtained as an integer combination of new basis elements.

Now the position vector of the centre of any face parallel to the plane defined by the vectors $\mathbf{a}$ and $\mathbf{b}$ is of the form

$$n\mathbf{a} + m\mathbf{b} + r\mathbf{c} + \mathbf{d},$$

which is an integer combination of the new basis vectors, namely

$$(n + m - r + 1)\mathbf{d} + (n - m + r)\mathbf{e} + (-n + m + r)\mathbf{f}.$$

Similarly for the centres of the faces in the other two directions.

The only task that remains is to check that the new basis does not generate any points *other* than original lattice points or points in the centres of faces of the old lattice. This is a bit trickier. One way is to note that, whatever the parities (even or odd) of n, m and r, the integers $n + m - r$, $n - m + r$ and $-n + m + r$ have the same parity. The original lattice points of L therefore correspond to integer combinations of $\mathbf{d}$, $\mathbf{e}$ and $\mathbf{f}$ whose coefficients have the same parity. Now *every* integer combination of $\mathbf{d}$, $\mathbf{e}$ and $\mathbf{f}$ is of the form $\mathbf{p}$ or $\mathbf{p} + \mathbf{d}$ or $\mathbf{p} + \mathbf{e}$ or $\mathbf{p} + \mathbf{f}$ where $\mathbf{p}$ is an integer combination whose coefficients have the same parity. Therefore, every lattice point of $L(\mathbf{d}, \mathbf{e}, \mathbf{f})$ is either a lattice point of L or a lattice point of L shifted by $\mathbf{d}$, $\mathbf{e}$ or $\mathbf{f}$. These three shifts correspond to moving to the centres of three faces.

Solution 3.4

We must show that the vectors $\mathbf{a}, \mathbf{b}, \mathbf{c}, \mathbf{a} - \mathbf{b}, \mathbf{a} - \mathbf{c}$ and $\mathbf{b} - \mathbf{c}$ are equal in length. Now,

$$\mathbf{a} - \mathbf{b} = \left(0, \tfrac{1}{2}, -\tfrac{1}{2}\right),$$
$$\mathbf{a} - \mathbf{c} = \left(\tfrac{1}{2}, 0, -\tfrac{1}{2}\right),$$
$$\mathbf{b} - \mathbf{c} = \left(\tfrac{1}{2}, -\tfrac{1}{2}, 0\right),$$

and so each of the six vectors has length $\tfrac{1}{\sqrt{2}}$.

Solution 3.5

Let us take $\mathbf{a} = (1, 0, 0)$, $\mathbf{b} = \left(\tfrac{1}{2}, \tfrac{\sqrt{3}}{2}, 0\right)$. The vector $\mathbf{c}$ has offset $\left(\tfrac{1}{3}, \tfrac{1}{3}\right)$ relative to $\{\mathbf{a}, \mathbf{b}\}$, and so

$$\mathbf{c} = \left(\tfrac{1}{2}, \tfrac{\sqrt{3}}{6}, c_3\right),$$

so that

$$\mathbf{c} \cdot \mathbf{c} = \tfrac{1}{4} + \tfrac{3}{36} + c_3^2,$$

and this must equal 1. The positive value of c_3 which satisfies this is the required height; it is $\sqrt{\tfrac{2}{3}}$.

Solution 4.1

(a) The symmetry group is still Γ(CUBE).

(b) The vertex type of the cube is (4,4,4).
The vertex type of the truncated cube is (3,8,8).
The vertex type of the cuboctahedron is (3,4,3,4).

Solution 4.2

(a) The symmetry group is still Γ(CUBE).

(b) The vertex type is (4,6,6).

Solution 4.3

The vertex type is (3,6,6).

Solution 4.4

The vertex type of the truncated dodecahedron is (3,10,10); that of the truncated icosahedron is (5,6,6); and that of the icosidodecahedron is (3,5,3,5). The symmetry group in each case is Γ(DODECA) = Γ(ICOSA).

Solution 4.5

The cube is a 4-prism and the octahedron is a 3-antiprism.

Solution 4.6

(a) $x_1^2 x_2^6$

(b) x_2^7

(c) $x_1^2 x_3^4$

Solution 4.7

Each rotation of order 3 about an axis through a vertex and an opposite face gives two 3-cycles of edges, and there are eight such rotations.

Each rotation of order 2 about an axis through a pair of opposite edges gives two 1-cycles and two 2-cycles, and there are three such rotations.

Remembering to account for the identity, we therefore obtain

$$P_G(x_1, x_2, x_3) = \tfrac{1}{12}\left(x_1^6 + 3x_1^2 x_2^2 + 8x_3^2\right).$$

The number of equivalence classes of colourings using three colours is therefore

$$\tfrac{1}{12}(729 + 243 + 72) = 87.$$

Solution 4.8

The rotations of order 3 give two 1-cycles and two 3-cycles, while the rotations of order 2 give four 2-cycles. Therefore

$$P_G(x_1, x_2, x_3) = \tfrac{1}{12}\left(x_1^8 + 8x_1^2 x_3^2 + 3x_2^4\right).$$

The number of equivalence classes of colourings using m colours is thus

$$\tfrac{1}{12}\left(m^8 + 11m^4\right).$$

Solution 4.9

Let us assume that the four faces of the cube that are divided by parallel lines are positioned vertically, the other two being horizontal (as in Figure 4.11). These vertical faces must be mapped to each other by any rotational symmetry, so the rotation group is a subgroup of $\Gamma^+(\mathrm{DIH}_4)$. But the divisions on the horizontal faces reduce the group still further to $\Gamma^+(\mathrm{DIH}_2)$.

The identity gives twelve 1-cycles and each of the rotations of order 2 gives six 2-cycles. Thus,

$$P_G(x_1, x_2) = \tfrac{1}{4}(x_1^{12} + 3x_2^6),$$

and the number of equivalence classes of colourings using two colours is

$$\tfrac{1}{4}(4096 + 192) = 1072.$$

Solution 5.1

(a) $r[\pi]\, q[(1,-1),(1,1),3\pi/4] = r[\pi]\, t[(1,-1) + 2(1,1)]\, q[3\pi/4]$
$ = r[\pi]\, t[(3,1)]\, q[3\pi/4]$
$ = t[(-3,-1)]\, r[\pi]\, q[3\pi/4]$
$ = t[(-3,-1)]\, q[\pi/4].$

(b) A glide reflection.

Solution 5.2

1 to 2: $g_{\frac{1}{2}}$; 1 to 3: $tg_{\frac{1}{2}}v$.

Solution 5.3

(a) $\frac{1}{10}(x_1^5 + 5x_1 x_2^2 + 4x_5)$

(b) $\frac{1}{10}\left((R+G)^5 + 5(R+G)(R^2+G^2)^2 + 4(R^5+G^5)\right)$

Solution 5.4

(a) $a\ \overrightarrow{b}\ \overleftarrow{b}$
$a\ \overrightarrow{b}\ \overleftarrow{b}$

(b) $n_t(\mathcal{T}) = 6$.

(c) By the Vertex Orbit Theorem,

$$n_v(\mathcal{T}) = 6\left(\tfrac{1}{6} + \tfrac{1}{6} + \tfrac{1}{6}\right)$$
$$= 3.$$

Solution 5.5

(a) $G_1 \cap G_2 = \{e, q[\pi/6]\}$.

(b) One possible rotation is $r[\mathbf{a} - \mathbf{b}, \pi/3]$.

One possible glide reflection is $q\!\left[-\tfrac{3}{2}\mathbf{a}, \tfrac{1}{2}\mathbf{b}, 0\right]$.

Solution 5.6

(a) Five.

(b) Either there are indirect symmetries or there are not. If not, there is one possibility, namely $p2$. If so, then there are axes in two directions, and as they are at right angles they are both rectangular or both rhombic. The former possibility gives three types (reflection axes in both directions, glide axes in both directions, or reflection in one and glide in the other). The latter possibility gives just one type, with alternating reflection and glide axes in both directions.

OBJECTIVES

After you have studied this unit, you should be able to:

(a) generalize to three dimensions the concept of a lattice and its basic properties;

(b) explain the classification of crystal patterns into seven crystal systems in terms of their point groups;

(c) identify the fourteen types of Bravais lattices, and identify the type of a space lattice given in terms of a basis in layer form;

(d) explain the construction of certain semi-regular polyhedra by truncating the regular solids;

(e) describe various ways of filling space with certain combinations of regular solids and semi-regular polyhedra;

(f) count the equivalence classes of colourings of various geometric configurations based on regular solids and semi-regular polyhedra.

INDEX

antiprisms 32
Archimedean solids 28
base-centred monoclinic lattice 23
base-centred orthorhombic lattice 18
basic parallelepiped 10
body-centred cubic lattice 8, 25
body-centred orthorhombic lattice 24
body-centred tetragonal lattice 25
Bravais lattices 7
crystal pattern 7
cubic crystal system 20
cuboctahedron 29
dual polyhedra 32
face-centred cubic lattice 8, 27
face-centred orthorhombic lattice 27
hexagonal crystal system 20

hexagonal lattice 8, 20
hexagonal space lattice 20
icosidodecahedron 30
identity of lattices 9
lattice 6
monoclinic crystal system 17
offset (three vectors) 14
offset (two vectors) 12
orthorhombic crystal system 18
polyhedron 28
primitive cubic lattice 20
primitive monoclinic lattice 17
primitive orthorhombic lattice 18
primitive tetragonal lattice 20
prisms 32
rhombic dodecahedron 33

rhombidodecahedron 33
semi-regular polyhedra 28
space lattice 6
tetragonal crystal system 20
three-dimensional 6
triclinic crystal system 21
triclinic lattice 21
trigonal crystal system 22
trigonal lattice 22
truncated cube 28
truncated dodecahedron 30
truncated icosahedron 30
truncated octahedron 29
truncated tetrahedron 30